S. E. Daniel, F. F. Cruz-Sánchez,
A. J. Lees (eds.)

Dementia in Parkinsonism

SpringerWienNewYork

Dr. S. E. Daniel
Dr. A. J. Lees
Parkinson's Disease Society Brain Research Centre (Brain Bank),
Institute of Neurology, London, United Kingdom
Dr. F. F. Cruz-Sánchez
Neurological Tissue Bank, Hospital Clinic, University of Barcelona, Spain

© 1997 Springer-Verlag/Wien
Softcover reprint of the hardcover 1st edition 1997

Product Liability: The publisher can give no guarantee for information about drug dosage and application thereof contained in this book. In every individual case the respective user must check its accuracy by consulting other pharmaceutical literature. The use of registered names, trademarks, etc. in this publication does not imply, even in the absence of a specific statement, that such names are exempt from the relevant protective laws and regulations and therefore free for general use.

Printed on acid-free and chlorine-free bleached paper

With 29 Figures

ISBN-13: 978-3-211-82959-2   e-ISBN-13: 978-3-7091-6846-2
DOI: 10.1007/978-3-7091-6846-2

Preface

In his original essay on the shaking palsy (1817) James Parkinson remarked that the "senses and intellect were uninjured". Thus, it was only in later years that the complexity of parkinsonism and in particular Parkinson's disease with dementia was recognised. Cognitive impairment in Parkinson's disease is common and is estimated to affect more than forty per cent of patients with disease onset after age 65. Recent studies suggest that pathology of Parkinson's disease now ranks second to Alzheimer's disease as the commonest substrate of dementia in elderly patients. The condition is heterogeneous and there remain many complicated and unresolved questions concerning cause, diagnosis and classification. In an attempt to clarify these issues, scientists and members of the European Brain Bank Network (EBBN) gathered in London for a meeting entitled "Dementia in Parkinsonism". The resultant monograph is testimony to the wide-ranging clinical, morphological and biochemical aspects of this condition. We are grateful to all contributors for expressing their expert opinions and for being so generous with their time taken in preparation of the manuscripts. The meeting was funded by the Commission of the European Communities as part of a Biomed-1 Programme. We wish to thank Amgen Limited, Lilly Industries Limited and Roche Products Limited for additional sponsorship. The expert secretarial assistance of R. Nani in the preparation of this book has been very much appreciated.

London, November 1997

S. E. Daniel
F. F. Cruz-Sánchez
A. J. Lees

Contents

Listed in Current Contents/Life Sciences

The language of the basal ganglia

G. Stern

Department of Clinical Neurology, University College London, United Kingdom

Summary. A brief historical account of the evolution of nomenclature frequently used in discussing the structure and function of the basal ganglia in health and disease is presented.

Introduction

During the course of this meeting I anticipate that frequent reference will be made to certain descriptive terms and concepts concerning the basal ganglia. I hope it will be of interest to the several contributors to be reminded of the origin of this "language". The topic has received the attention of many distinguished scholars and I must confess that I have flagrantly plagiarised some of their contributions in presenting the following brief account.

Deciphering ancient origins and etymology requires an intimate knowledge of classical Greek, Syriac, Arabic and mediaeval Latin with which few of us are blessed; additional problems stem from the fact that mediaeval anatomists had no access to the original Greek manuscripts of Hippocrates and his school which have not survived in their original form, but only through several translations. Thus the Latin versions were based upon the texts of Arabic scholars, particularly Avicenna and Hainan written in the ninth and tenth centuries; accurate understanding was further obfuscated by the fact that Arab physicians were unable to read directly Greek texts having only the writings of an intermediary Syriac culture. Understandably many errors may have occurred over the centuries when sequentially Greek, Syriac, Arabic, Latin and then contemporary English and European languages were used in the course of evolving translations. A further and probably considerable source of error until the sixteenth century was the absence of surviving drawings and diagrams and for the inheritors of this tradition, total dependence on written descriptions.

For example, Galen clearly had detailed knowledge of the ventricular system of the brain and adjacent structures, having acquired his knowledge of anatomy from Herophilus who had first routinely performed human autopsies in the Alexandrian school of medicine around 300 B.C. Thus, Galen was clearly aware of the ganglia at the base of the lateral ventricles and described them as gluteal parts, elsewhere stressing their visual similarity to the thighs.

However, Lewy speculated whether the term "caudate nucleus" evolved through an error in translation. He pointed out that the Greek synonym of "glutia" is similar to the word for tail and suggested that Avicenna thus translated the term without justification from the Greek text such that others derived the nucleus "caudatus" as the tailed nucleus.

Further confusion was caused by prevailing views relating structure to function and at times mixing both categories. In the "Anathomia" of Mundinus (1538) which influenced European medical terminology for over two centuries and formed the foundations of modern medicine and terminology even to this day, an exotic mixture of Greek, Roman, Arabian and mediaeval philosophy can be discerned. Thus, Mundinus "at the base of the lateral ventricles lies a red sanguineous substances called vermis . . . because it looks like an earthworm and because it behaves like it . . . namely it can contract . . . and in doing so blocks the communication from one ventricle to the other . . . (prevailing philosophy located fantasy in the anterior ventricle, imagination in the medial and memory in the posterior ventricle) . . . if then the individual wishes to suspend cogitation, he elongates his worms and this prevents fantasy, memory and imagination from uniting and forming the 'common sense' . . .". It is also difficult to discern the slow evolution of terms from that used by our ancient predecessors to present designations for the basal ganglia. For example, among the first illustrations of the open brain drawn from nature by Étienne (1545) old names are still employed in an illustration which depicts the ventricular system with reasonable accuracy.

We owe most of our present nomenclature to the English physician Thomas Willis. He had the good fortune to have as his artist and illustrator Sir Christopher Wren and "Cerebri Anatome" (1664) clearly illustrates the naked eye appearance of the cut brain. ". . . in the foremost part of the brain stem two prominent lentiform bodies . . . which are characterised by up and downward radiating mediated fibres producing a striped appearance on their cut surface . . .". Willis used the terms lenticular and striatal bodies synonymously and described this structure as "dissimilar to any other part of the brain".

Willis quotes an Arabic text concerning the glutia and identifies striatum with glutia. He appears to be the first observer to clearly differentiate the striped anterior protuberance from the posterior non-striated lobe for which he introduced the term "thalamus opticus". The term probably arose from another error of translation as Lewy — vide infra — drew to our attention. The Greek "thalamos" means a room or cavity and was used in this sense by Galen to describe the chambers of the heart. The recess of the inferior horn of the lateral ventricle at its upward bend is spoken of by Galen as thalame, the loophole which receives the "spirit" by way of optic nerves which were thought to be hollow. Hainan, Galen's Arabic interpreter, likened this region with its broad base and apex — using the time-honoured system of conveying the description of an anatomical structure by comparison with a commonplace and widely recognised object — as a "sleeping tent". Willis appears to have been more impressed by "sleeping" than by "tent" observed Lewy and thought more of a bed than a bedroom, confusing, but not too surprising, equivocal Greek and Arabic words for the same object. Whatever his intentions, it seems that Willis compromised between the two quotations changing "thalame" into "thalamos" and applied the name to the ganglion instead of the adjacent ventricle — and thus emerged "thalamos nervorum opticorum" — subsequently abbreviated to thalamos opticus and presently to thalamus. Thus from the mists of history emerged the terms corpus striatum, nucleus caudatus (caudate), and thalamus. Before leaving Willis's seminal contributions to neuroanatomy his remarkable prescience concerning function may be noted. "The corpus striatum represents an exchange between brain stem and cortex . . . the corpus striatum receives the notion of spontaneous localised movements in ascending tracts . . . conversely, from here tendencies are despatched to enact motions without reflection over descending pathways" (1664).

The term basal ganglia gained gradual acceptance towards the end of the nineteenth century and Ringer (1879) may have been the first when he wrote "athetosis is due to atrophy and degeneration of the basal ganglia"; his description indicates that he included the putamen and globus pallidus. Gowers (1885) included putamen, pallidum and thalamus under "basal ganglia" in his first edition of "Lectures on Diagnosis of Disease of the Brain" as well as in his celebrated Manual, but he also spoke of "central ganglia"; basal ganglia had yet to be generally accepted. For example, as late as 1896 Kölliker included "the basal ganglia of Meynert" and others preferred older titles such as the anterior and posterior cerebral ganglia or the grey ganglia. There were also territorial differences: some wished to include the corpus striatum, globus pallidus, optic thalamus, habenular ganglion, the putamen and globus pallidus.

The substantia nigra

Later contributions to the language of the basal ganglia are of course better documented and more accessible. The earliest description of the substantia nigra is generally attributed to Samuel Thomas von Sömmering (1755–1830) who was born at Thorn in Poland. His father, a physician, early instilled in his son an interest in medicine and frequently took him to autopsies. Sömmering entered the University of Göttingen in 1774 and in 1778 presented his doctoral thesis "De basi encephali et originibus nervorum". This remarkable paper described and classified the cranial nerves superseding Willis's description and contains the first accurate account of the structures of the base of the brain. Sömmering also drew attention to the difference in appearance between the ashen or grey matter (substantiae cinereae) of the cortex and the brain stem. "The mass is tinged a dark colour which in adults resembles neither the whiteness of the medulla nor the cinereal parts of the brain, but is so to speak, midway between the cinereal and medullary parts" and later described a particular aggregation of dark substance within the cerebral peduncles.

Sömmering was familiar with the intimate relationship of this pigmented structure to the emerging third nerve fibres and observed that pigmentation was less distinct in the brain of newborn children and foetuses. While Sömmering remained faithful to Galenical doctrines in placing the soul in the cerebral ventricles there can be little doubt about the originality of his anatomical observations.

Alexander Munro (1783) did not seem aware of Sömmering's priority. "Besides the cineritious matter observed in the cortical parts of the brain and cerebellum or on the surface of the corpus striata, a great deal of it is to be found enclosed within the medulla. Nay, in the middle substance of the brain and cerebellum, halfway between the surface of these and their ventricles and even within the cura cerebri, or the tuber annulare and medulla oblongata which are generally considered as pure medullary cords, I have found a great quantity of ciniterous matter". Three years later Vicq d'Azyr (1786) while acknowledging Sömmering's original description of a particular cerebral convolution failed to mention his priority concerning the substantia nigra. "I call this area, tâche noir or locus niger crurum cerebri". Despite these later observations most anatomists equate "Sömmering's substance" with the substantia nigra and while early nineteenth century observers including Bell (1823), Cloquet (1828) and Foville (1844) were familiar with its gross structure, it was much later before the substantia nigra was accepted to be part of the language of the basal ganglia. The findings of Blocq and Marinesco (1893) and of Trétiakoff (1919) were among the first to relate the substantia nigra to disordered movement.

The subthalamic nucleus

This nucleus has enjoyed a number of synonyms, including corps de Forel by Edinger, discus lentiformis by Meynert and nucleus amygdaliformis by Stilling, but most authors associated the name of Luys to the discrete, visually discernible nucleus of the subthalamus which Luys showed in his woodcut-illustrated textbook; Luys modestly but erroneously called the nucleus "bandelette accessoire de l'olive supérieure".

As the nucleus of Luys currently enjoys a notable resurrection of interest by those concerned in movement disorders it may be of interest to outline three distinct eras of research activity. Post-hemiplegic involuntary movements were known to Hammond (1871), Gowers (1876), Charcot (1879) as well as to others. Soon afterwards a second stage of clinico-pathological correlation evolved from several reports from different countries over the period 1897–1928 implicating the Corpus Luysii with dyskinesias. Amongst these were Bonhoeffer (1897), Touche (1901), Sano (1910), Jakob (1923), Martin (1927) and Von Santha (1928).

The next critical phase was the discovery by Whittier and Mettler (1949) that experimental electrolytic lesions in the monkey's subthalamic nucleus consistently provoked contralateral dyskinesia and over the next decade the anatomical, physiological and pharmacological implications were fully exploited as far as then available techniques permitted.

There was then an interlude of about two decades before the present wave of renewed interest commenced. DeLong and his colleagues initiated a further series of anatomical studies including detailed electrophysiological observations, somatosensory mapping, microstimulation, reversal of experimental parkinsonism by subthalamic lesions and parallel changes in motor behaviour and neuronal activity in the internal pallidum. These cumulative studies clarifying the role of the subthalamic nucleus in disordered movements and parkinsonism provided the background to the most recent experimental — and possibly therapeutic — development, namely bilateral stimulation of the subthalamic nucleus in patients with Parkinson's disease and the beneficial effect upon symptoms and signs (Limousin et al., 1995).

Lewy bodies

Frederic Lewy (1885–1950) was born in Berlin, the son of a distinguished physician and whose family included Paul Ehrlich. Lewy's teachers included Oppenheim, Cassirer, Krapelin, von Monakow, Magnus, Nissl, Alzheimer and Spielmeyer and from 1930–1933 he was professor of neurology and a director of the Berlin Neurological Institute. While working in Alzheimer's neuropathology laboratory in Munich he began to study serial sections from the brains of those with "paralysis agitans" and first described his seminal findings in 1910 reporting similar conclusions in 1913. He drew attention to characteristic globoid and elongated inclusions with a surrounding halo in nerve cells of the dorsum nucleus of the vagus and in Meynert's nucleus basalis summarising his considerable experience in a monograph "Muscle tone and movement" (1923). ". . . these changes can be characterised as deposits . . . features of ball-cord or serpent-like forms which are stained in a shining red and lying in a blue mass of plasma . . . there are no neuronal nuclei in these structures . . . it is also possible that some of these masses without a nucleus had earlier been ganglion cells with a nucleus . . . the list of chemical reactions are quite typical . . . globoid or oblong condensation of plasma which yield some reactions similar to corpora amylacea but which can also be stained by eosin. The plasma shows vacuoles, the core becomes detached and finally only these globoid forms . . . remain lying in the tissue. A glial scar ends the process . . . glassy degeneration or fuchsin-stained granules . . .". It was clear that Lewy regarded these changes as a dynamic process which in sections might be seen in different stages of development. These are clearly demonstrated in a series of drawings and it is of interest that whereas Lewy was impressed with the frequency of these findings in the dorsal motor nucleus of the vagus (a finding which would have pleased James Parkinson as he had speculated that the disease locus lay therein), in Meynert's nucleus, the nucleus lateralis thalami and paraventricular nucleus, he was not impressed by the extent of involvement of the substantia nigra. Thus in one of his series he found involvement in only eleven of fifty patients.

His name was soon generally accepted to be associated with these characteristic eosinophilic bodies and it seems that Trétiakoff (1919) was the first to use the term "corps de Lewy" for such changes in the substantia nigra and thereafter Lewy bodies became part of the language of the basal ganglia.

Apoptosis

The cryptic nature of the selective vulnerability of certain parts of the central nervous system associated with neurodegenerative diseases in the absence of any evident clinical or neuropathological explanation has long occupied the thoughts of neuroscientists. Explanations have remained conjectural but there has clearly been a need for a specific "language" or neologism to succinctly convey the essence of the process.

For example, Gowers (1899) when considering the nature of the focal primary degeneration of nerve elements in tabes dorsalis suggested that "differences in nerve element function involve variations in their susceptibility . . .". He returned to this theme in 1902 ". . . many of these parts have their own vitality . . . some of them may slowly die, while the life of all the rest goes on without impairment . . . we speak of it as 'degeneration', but the process is in many cases, perhaps in most, an essential failure of vitality . . .". He had long maintained "that this was the essential secret of the mysterious incidence of degenerative diseases . . . an average man inherited an equal vitality in all his structures. He went to pieces at last, like the famous 'one-hoss shay' all at once, in mature, equable senile decay . . . in other words a defect in vital endurance was only sufficient to render some structures less resistant than others . . . this disposition to decay, sometimes inherent, strongly or slightly, seemed to be induced artificially by toxic or perhaps other influences. The influence was specific in many cases probably more than they could discern and led to failure of nutrition, to decay and degeneration . . . mental change, especially simple mental failure, also often occurs under the same conditions and no doubt from a slow degeneration of cerebral neurons which connect to combine others in a way we cannot yet perceive. Another senile malady, paralysis agitans must be referred to vital failure in some cerebral motor structure . . . it generally occurs without adequate cause and must be regarded as a special defect in vitality . . .".

Here Gowers explains his introduction of the term abiotrophy. "I met with a difficulty, we have no word by which to designate this conception — I do not like new words — indeed I dislike them — but if we have a conception for which no name exists which we need frequently to speak of it is not wise, I think, to shrink from an attempt to give it a name. Here the simplest mode of obtaining what we need is to insert the root of 'bios' after the negative particle in 'atrophy' which gives us 'abiotrophy' . . . the adjective biotic has been occasionally used in English (first in 1600) . . . it exists in French as biotique from the Latin bioticus and the Greek biotos has the sense of 'lifeless'. Thus 'abiotic' is clearly legitimate . . . the word biosis was employed . . . in the old Greek meaning 'mode of life'. It is given us by the physician who is more widely known and esteemed on account of his writings than any other medical man who lived — St. Luke. It is pleasant I think to take a word from him . . .".

The term abiotrophy failed to survive into current usage, but this was also the fate of others who wrestled with the same problem. Thus Vogt (1925) described how he and his wife Cécile proposed the term "pathoklise" to convey the concept of selective vulnerability within the nervous system and in a siminal paper "Der Begriff der Pathoklise" attempts to dissect this in terms of the then recognised pathogenic mechanisms. Other than suggesting it was a subtle chemical-physical interplay of factors that remain to be elucidated he felt unable to advance his proposition further. Pathoclisis too fell out of common usage.

Abiotrophy, pathoclisis and other synonyms have recently undergone intellectual resurrection emerging as the concept of "apoptosis" now synonymous with programmed cell death. These terms are now widely used in the language of the basal ganglia and are certainly the most fashionable — at least judged by the volume of papers presently generated. Like the evolution of many concepts its development has not been smooth. In 1972 Kerr, Wyllie and Curry proposed the term to describe "hitherto little recognised mechanism of controlled cell deletion which appears to play a complementary but opposite role to mitosis in the regulation of animal cell populations". They described a characteristic sequence of structural changes of which the main elements were nuclear and cytoplasmic condensation and fragmentation of the cell into a number of membrane-bound, ultrastructurally well-preserved fragments followed by a second stage when "apoptic bodies are shed or taken up by other cells within phagosomes and rapidly degraded". The authors speculated that there might be inherently programmed events determined by intrinsic clocks specific for the cell type which provided a possible explanation for hitherto inexplicable cell death mechanisms for neurons as well as other cells. There was a considerable delay before the implications of a structurally distinct programmed cell death pathway for neurodegenerative diseases was perceived and exploited. At present it is a growth industry. Some of the reasons for this relatively late but now crescendo research development may have been the delay before the techniques of molecular biology were applicable and the availability of the increasing family of nerve growth factors; perhaps another reason was the more recent availability of amenable animal models. Whether or not there is substance to the apoptosis hypothesis the term has clearly passed into the language of the basal ganglia and it seems legitimate to assume that it will be used — alongside caudate, striatum, thalamus, substantia nigra, subthalamus and Lewy bodies — in the course of your further deliberations.

References

Bell J, Bell C (1823) Anatomy and physiology of the human body

Blocq P, Marinesco G (1893) Sur un cas de tremblement Parkinsonien hémiplegique symptomatique d'une tumeur du peduncule cérébrale. Cr Soc Biol Paris 5: 105

Bonhoeffer K (1897) Ein Beitrag zur Localisation der choreatischen Bewegungen. Monatsschr Psychiat Neurol 1: 6–41

Charcot JM (1897) Lectures on diseases of the nervous system (translated by Sigerson G). Henry C Lea, Philadelphia, p 21

Cloquet H (1828) A system of human anatomy. Masson, Paris

DeLong MR (1985) Primate globus pallidus and subthalamic nucleus: functional organisation. J Neurophysiol 53: 530–543

Étienne C (1545) De dissectione partium corporis humani libri tres a Carolo Stephano doctore medico editi. Una cum figuris et incisionum declarationibus a Stephano Riverio, chirurgo compositis. Simon Colinaeus, Paris

Foville AL (1844) Traite complet de l'Anatomie et de la Pathologie des Système Nerveux

Gowers WR (1876) On athetosis and posthemiplegic disorders of movement. Med chir Trans 59: 271

Gowers WR (1885) Lectures in the diagnosis of the diseases of the brain. Churchill, London

Gowers WR (1899) The pathology of tabes dorsalis and general paralysis of the insane. Pathological Society of London. Lancet: 1591–1596

Gowers WR (1902) A lecture on abiotrophy. Lancet April 12

Hammond WA (1871) A treatise on diseases of the nervous system. Appleton & Co, New York, p 654

Jakob A (1923) Die extrapyramidalen Erkrankungen. Springer, Berlin, p 328

Kerr JFR, Wyllie AH, Currie AR (1972) Apoptosis: a basic biological phenomenon with wide-ranging implications in tissue kinetics. Br J Cancer 68: 239–257

Kölliker A (1896) Handbuch der Gewebelehre, Bd 2, 6. Aufl. Engelmann, Leipzig, p 456

Lewy FH (1912) Paralysis agitans. I. Pathologische Anatomie. In: Handbuch der Neurologie III. Springer, Berlin, pp 920–933

Lewy FH (1913) Zur Pathologischen Anatomie der Paralysis Agitans. Dtsch Z Nervenheilk 50: 50–55

Lewy FH (1923) Die Lehre vom Tonus und der Bewegung. Springer, Berlin

Limousin P (1995) Effect of parkinsonian signs and symptoms of bilateral subthalamic nucleus stimulation. Lancet 354: 91–95

Luys J (1876) Le cerveau et ses functions. Librairie Germer Baillière et Cie, Paris

Martin JP (1927) Hemichorea resulting from a local lesion of the brain (syndrome of the body of Luys). Brain 50: 637–651

Mundinus (1538) Anathomia Mundini per Carpum castigata et postmodus cum apostillis ornata et noviter impresses. Bernardinus, Venice, pp 61, 73

Munro A (1783) Observations on the structure and function of the nervous system, ch 7. Edinburgh, p 23

Ringer S (1879) Notes of a post-mortem examination on a case of athetosis. Practitioner 23: 161

Sano T (1910) Beitrag zur vergleichenden Anatomie der Substantia nigra, des Corpus Luysii und der Zona incerta. Monatschr Psychiat Neur 27: 110–127

Sömmering ST (1778) De Basi Encephali et Originibus Nervorum. Göttingen

Touche R (1901) Deux cas d'hémichorée organique avec autopsie. Rev Neurol 9: 1080–1081

Trétiakoff C (1919) Contribution à l'étude de l'anatomie pathologique du locus niger der Sömmering avec quelques decutions relatives à la pathogenie des troubles du tonus musculaire et de la maladie de Parkinson. Thesis, University of Paris

Vicq d'Azyr (1786) Traite d'Anatomie et Physiologie, vol 1, ch XXII. Paris

Vogt O (1925) Der Begriff der Pathoklise. J Psychol Neurol 31: 245–255

Von Santha K (1928) Zur Klinik und Anatomie des Hemiballismus. Arch Psychia 84: 164–178

Whittier JR, Mettler FA (1949) Studies on the subthalamus of rhesus monkey; hyperkinesia and other physiologic effects of subthalamic lesions, with special reference to subthalamic nucleus of Luys. J Comp Neurol 90: 319

Willis T (1664) Cerebri Anatome cui accessift Nervorum Descriptio et usus (illustrated by Sir Christopher Wren). Roycroft, London

Author's address: G. Stern, M.D., The Woolavington Wing, Middlesex Hospital, Mortimer Street, London W1N 8AA, United Kingdom

Aging and the nigro-striatal pathway

F. F. Cruz-Sánchez, A. Cardozo, C. Castejón, E. Tolosa,
and **M. L. Rossi**

Neurological Tissue Bank, Hospital Clinic, University of Barcelona, Spain

Summary. Aging is associated with a progressive impairment in motor function. This feature, together with the decline in mental function, could be considered as an aging syndrome which may finally compromise the ability of the elderly to maintain an active, independent life-style.

In the present paper a wide variety of morphological aspects, which have been classically related to brain aging and others such as cytoskeletal changes, the role of growth factors and molecular changes, will be reviewed focusing on aging of the nigrostriatal pathway. In addition to sharing features of aging common to other structures, it is likely that the nigrostriatal pathway has specific characteristics derived from its particular molecular characteristics and/or from a selective vulnerability to aging.

To gain further insight into the aging syndrome, the acquisition of rigorous criteria for selecting control cases is paramount. The improvement of methods for the preservation of human tissue is also crucial.

Introduction

Aging is associated with a progressive impairment in motor function characterized by slowness of movement, stooped posture, and shuffling gait which could resemble, although to a lesser degree, some clinical features of Parkinson's disease (PD) (Terävainen et al., 1983). These motor changes together with the decline in mental function also associated with aging, could be considered as an aging syndrome which may finally compromise the ability of the elderly to maintain an active, independent life-style.

Aging has been considered to have an important role in pathophysiological mechanisms of degenerative processes such as PD, Alzheimer's disease (AD) and amyotrophic lateral sclerosis. These mechanisms have been linked to a selective vulnerability of neurons and to the presence of different morphological hallmarks (Lewy bodies, senile plaques, Bunina bodies) which have been called the "gravestones" of neurons (Calne et al., 1986). However, the emphasis on aging as a contributory factor to the pathophysiology of these disorders declined with the increasing characterization of degenerative processes. Nowadays, it is even possible to detect early neuropathological stages of the degenerative process in asymptomatic patients (Braak and Braak, 1991; Gibb and Lees, 1991).

In the present paper a variety of features which have been classically related to brain aging and others such as cytoskeletal changes, the role of growth factors and molecular changes will be reviewed. Furthermore, we will discuss the importance of control cases and methods for their selection to study morphological, biochemical and molecular parameters including:

1. Neuronal loss
2. Dendritic spines and synaptic changes
3. Cytoskeletal changes
4. Growth factors
5. Neuropigments
6. Plasticity of the nervous system

1. The definition of "normal" cases for controls

Selection of control cases for brain banking requires sampling of non-neuropsychiatric cases using the same protocol applied to some neurodegenerative diseases such as parkinsonian cases.

Essential information to be gathered includes (for comprehensive criteria, see Gsell et al., 1993):

Pre-mortem information

Age and sex.

— Clinical data:
  • Family history (dementia, intellectual impairment, movement disorders, psychoses, personality disorders).
  • General history (cardiovascular, metabolic, respiratory, haematological, neoplastic, hormonal, infections, etc.).
  • Psychiatric history (manic-depressive disorders, psychoses, psycho-reactive and neurotic disorders, personality disorders, suicide attempts, hallucinations, etc.).
  • Neurological history.
  • History from family or health carers (personality or mood changes, slowness of ideation, memory and concentration disturbances, speech impairment)
  • Medication history.
  • Laboratory tests.
  • Neuropsychological assessment.
  • General physical and neurological examination.
— Agonal status.

Given an equal post-mortem interval, and in the absence of a specific pathological condition, the best indicator reflecting the severity of the agonal status is the pH of the brain tissue and of the CSF (Ravid et al., 1995).

Post mortem information

— Date and time of death.
— Clinical cause of death.
— Post-mortem delay and condition of body storage.
— Pathological information: macroscopical examination of the brain and of the rest of the body.
— Microscopy: histological examination of the brain, spinal cord and routine sampling of other organs.

Central nervous system (CNS) diagnosis is based on the examination of multiple areas using routine staining techniques and immunohistochemistry where applicable. In our experience for the study of aging, a normal case should only be designated as a non-neuropsychiatric control after a thorough pathological examination has been made, as this may well result in the elimination of a high proportion of spurious cases. The most common CNS finding in such cases is hypoxic change, closely followed by changes associated with the pathology of microcirculation related to common conditions such as diabetes or hypertension. Most of these changes e.g., status lacunaris, lacunae or calcifications of the basal ganglia (Fig. 1A,B), may compromise the nigrostriatal pathway. Tissue rarefaction, gliosis and neuronal loss in striatal structures may also result in indirect damage to the substantia nigra. It is important to take into account that most aged patients may suffer from this clinically silent process, and exhaustive neuropathological studies may contribute to confirm the presence of similar changes (Fig. 2) and/or abiotrophic substantia nigra damage.

Degenerative changes involving several CNS structures may also be observed on detailed neuropathological examination e.g., neurofibrillary tangles (NFT), senile plaques (SPs), Hirano bodies. Although most of these changes have been repeatedly addressed in the literature as age-related, recent studies demonstrate that they can be the harbinger of neurodegenerative conditions such as AD (Braak and Braak, 1991) or PD (Gibb and Lees, 1991). This point will be further discussed below.

It follows that pathological findings in an elderly potential control case are not uncommon, whether or not there have been any clinical signs. For this reason, we suggest that many cases used as "controls", are likely to have relevant pathological conditions, which should be specified and quantified.

According to Ravid et al. (1995) the interpretation of human brain data remains extremely difficult due to the variety of confounding factors that may change the tissue and therefore need to be excluded. In conclusion, gathering "normal" controls must mean that full pathological examination is performed in addition to spending the time collecting full clinical information.

2. Neuronal loss

The concept of neuronal loss with aging was first proposed by Hodge (1894). There is no doubt that direct morphometrical studies provide the best evi-

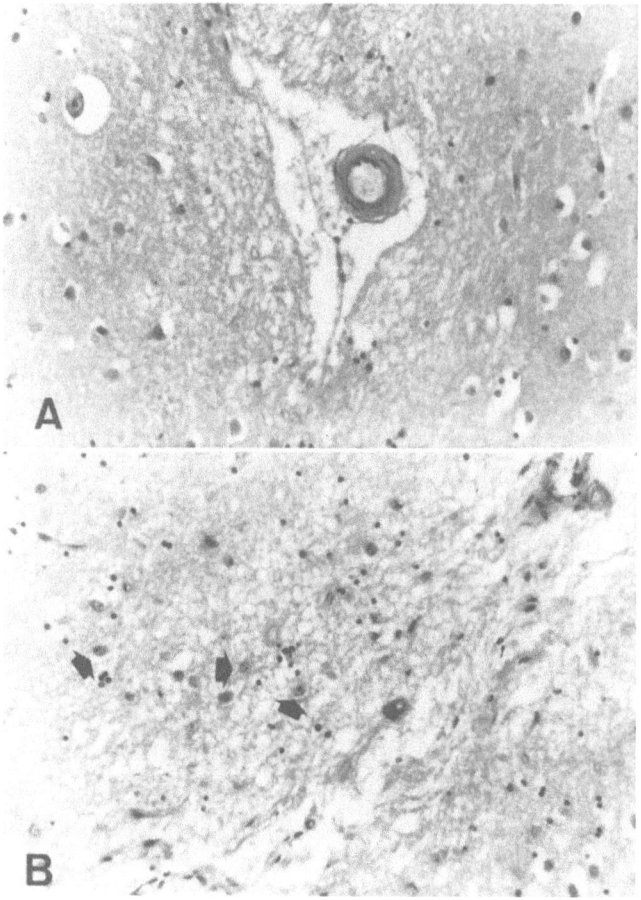

Fig. 1. Status lacunaris in the putamen of a non-neuropsychiatric man, 83 years old (**A**). Note the fibrohyalinosis of the central vessel (H&E ×250). Micronecrosis in the putamen of a non-neuropsychiatric man, 78 years old (**B**). Observe tissue rarefaction and proliferation of hypertrophic astrocytes (arrows), (H&E ×150)

dence of the rate of striatonigral attrition as witnessed by several studies (Mc Geer et al., 1977; Mann et al., 1984; Calne and Peppard, 1987; Thiessen et al., 1990; Fearnley and Lees, 1991). The main interest in this line of research is to look for a possible contribution of the aging factor to the pathogenesis of PD. However, substantia nigra neuronal loss with aging has been a lesser focus of interest since knowledge has far advanced in the field of neurodegeneration.

McGeer et al. (1977) found a fall in the number of nigral neurons which was directly proportional to age with 48% loss by the age of 60, (about 7% per decade), in a series of 28 cases without neurological disease. Although the methodological approach is not fully described, this work remains an important point of reference to support the concept of nigral neuronal loss with aging.

Mann et al. (1984) studied the number of substantia nigra neurons at the level of origin of the third nerve in 67 "controls" (aged from 11 to 97 years). They found 35% neuronal loss by 90 years of age. Thiessen et al. (1990)

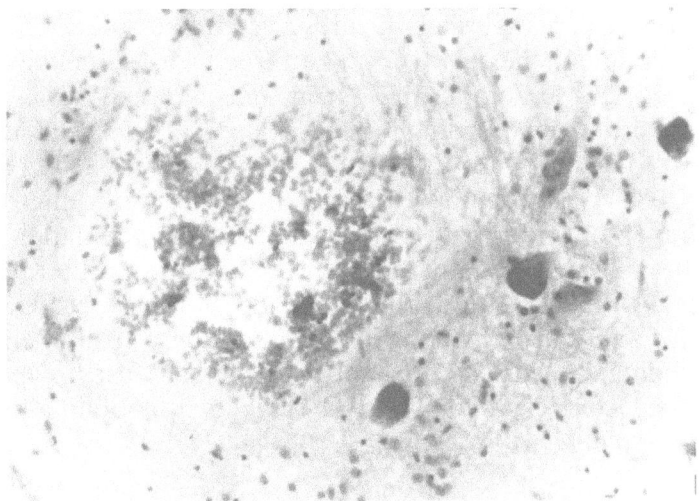

Fig. 2. Microhaemorrhage in the substantia nigra of a non-neuropsychiatric woman, 74 years old (H&E ×350)

quantified nigral neurons at the same level from the substantia nigra in 22 rural and 12 urban control cases and found a greater decline (pigmented and non-pigmented cells) in the urban population (34% vs. 17%) a contribution from environmental factors. Information regarding type of control cases selected for this research was not specified. Fearnley and Lees (1991), in a thorough study of 36 control cases aged 21 to 91, found a linear 33% fallout of substantia nigra neurons with aging. A subdivision of the pars compacta into ventral and dorsal with reference to the fibres of the third nerve was made. A 6.9% neuronal dropout per decade was observed in the dorsal part.

It is evident that all these studies are in agreement concerning the loss of neurons in aging, but findings in relation to the degree of cell loss are quite variable and cases selected as controls may not have been free of other types of pathologies such as those mentioned above in section 1.

In an attempt to confirm and quantify neuronal loss in the substantia nigra with aging, 40 "normal" cases out of 100 potential controls were selected with the remainder excluded for various reasons according to pre- and post-mortem data extensively discussed in section 1.

Causes for exclusion were pre-mortem factors in 24% including agonal state 6%, post-mortem delay 14%, and histological features of ischaemic or degenerative changes (senile plaques, neurofibrillary tangles and Lewy bodies) in 22%.

In all cases studied the substantia nigra was separated from the brain stem by cutting at right angles to its long axis, from the lower border of the superior colliculus to the point of origin of the third cranial nerve. A block about 15 μm thick was obtained and divided parasagittally into two parts. At the level of the third cranial nerve, 10 serial 20 μm thick sections were stained with H&E. Three sections from this level, the first, the fifth, and the tenth were used for morphometric study. An amplified image of pigmented neurons was projected from a microscope (original magnification ×140) to a screen at a distance of 1.20 µm to be counted, and the average of three measurements was obtained as representative of the number of neurons in each case.

Cases were grouped according to age: group A (17–59 years old), group B (60–93 years old). Significant differences were found in older cases, with a decrease of 36% in the number of pigmented substantia nigra neurons in group B compared with group A. Results showed a smaller decrease of neurons with aging when compared with those from the literature. Criteria used to obtain a qualified and homogeneous "normal" group may be the reason for the difference between our results. These results emphasise the value of pathological examination and the rigorous use of morphological criteria when selecting cases for the study of aging, to obtain a homogeneous disease-free sample.

3. Dendritic, spines and synaptic changes in aging

Several authors have described morphological changes associated with aging in various CNS regions with the Golgi method. This technique is valuable in demonstrating cytoarchitecture and is one of the best and most elegant

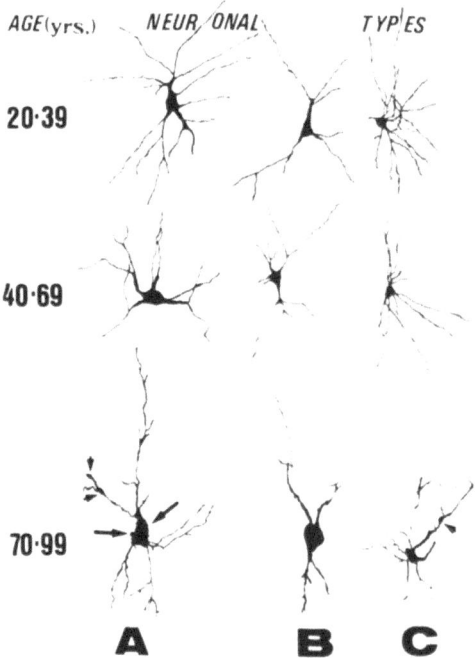

Fig. 3. Microdrawing of different neuronal types of human substantia nigra (A: type I, B: type II, C: type III). Cells are grouped according to age. Note that neurons from the oldest group show swelling and distortions (arrows). Dendrites are fewer, shorter, and have some nodulations (arrowheads). Changes are conspicuous in cell types I and II and less evident in cell type III (Camera lucida ×60) (reproduced with permission from J Neuropathol Exp Neurol)

ways to provide clarity and delineation of the shape and spatial arrangement of axons and dendrites (Valverde, 1970; Scheibel, 1970; Ramón-Moliner, 1970). Many authors have described morphological abnormalities with aging in cortical and hippocampal neurons with this technique (Scheibel, 1976, 1977). These changes consisted of distortion of the cell body profile and dendrites. Machado-Salas et al. (1977) described similar changes in the spinal cord and lower brainstem of the aging mouse. Cruz-Sanchez et al. (1995) studied human aging changes in silver impregnated substantia nigra neurons. Three types of neurons were found according to shape and size: type I (the largest ones in the pars compacta), type II (medium size, in the pars reticularis) and type III (the smaller ones, in the pars reticularis). Findings observed during aging in these neurons are distortion of cell body profile and loss of spines and dendrites (Fig. 3) which is more pronounced the greater the

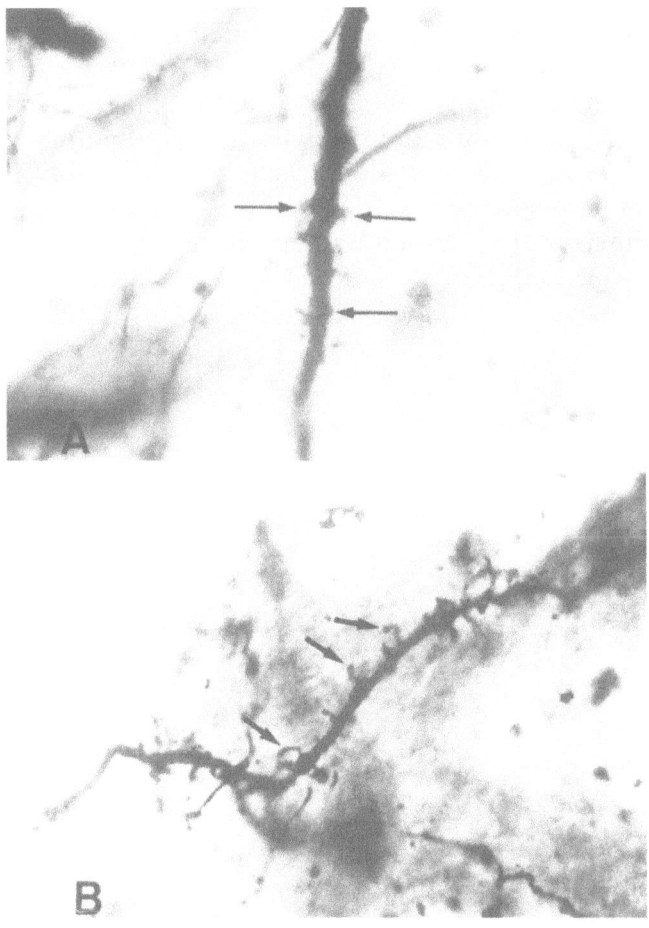

Fig. 4. Dendrite of a substantia nigra type I neuron from a young man, 27 years old (**A**), showing a homogeneous profile of spines (arrows) compared with a dendrite of the same type of a neuron from an 85-year-old man (**B**). In this case spines are thin, elongated and distorted (arrows). (Rapid Golgi method, ×400) (reproduced with permission from J Neuropathol Exp Neurol)

distance from the cell body (Fig. 4A,B), and swelling and beading of the remaining dendrites. These changes were observed more frequently in the aged group (from 70 to 93 years) (Fig. 5A,B). In spite of common findings to those described before, we found that different types of neurons were compromised to a varying extent with a predominant involvement of type I neurons (Fig. 5A,B). On the other hand, cytoskeletal abnormalities could not be demonstrated in an immunohistological study of samples from these series. Some of the morphological findings found, namely nodulations and beading of dendrites, showed similarities with those observed earlier in mice substantia nigra neurons treated with methyl-4-phenylpiridine (MPTP), but cytoskeletal changes were not encountered (Cruz-Sanchez et al., 1993; Boatell et al., 1992). Patt et al. (1991) also described dendritic changes in type I substantia nigra neurons in PD, however, these were attributed to Lewy bodies. The pathophysiological mechanism underlying morphological changes observed in silver impregnated aged humans substantia nigra sections remains unknown, but it may be akin to MPTP toxicity.

Synaptophysin (Sph), an abundant membrane protein of presynaptic vesicles has been shown to be a useful marker for synapses (Wiedenmann, 1985). In an attempt to detect possible synaptic changes of the nigrostriatal pathway with aging, Sph immunoreactivity of nigral neurons was studied in our series of control cases and slight reactivity was observed in the oldest cases. This Sph immunoreactivity decrease may confirm the compromise of dendritic terminals at the substantia nigra during aging demonstrated previously in the Golgi study.

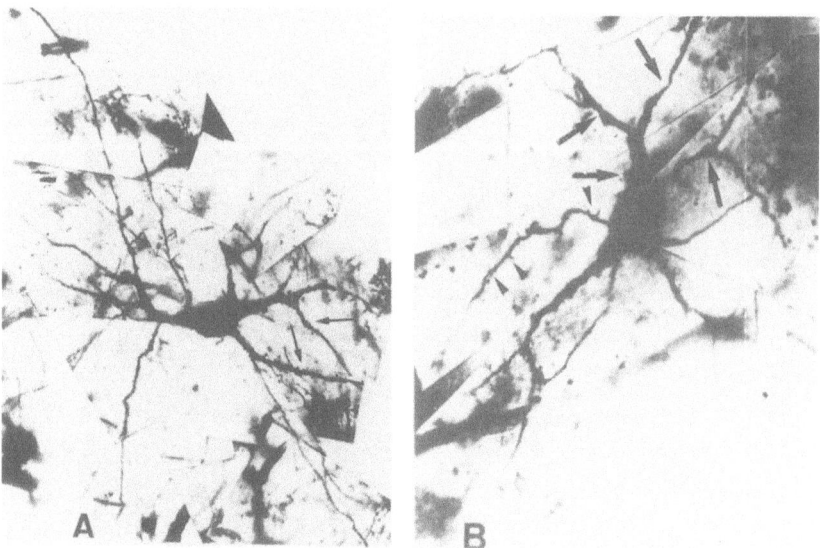

Fig. 5. Substantia nigra type I neurons from a 27 year old (**A**) and an 83-year-old (**B**) man. Compare the cell body and dendritic profiles as well as thickness and length of dendrites. Distorted profile of the cell body and dendrites (arrows) are conspicuous in B. The number of spines is also reduced in B (arrowheads), compared with the homogeneous profile of spines in A (arrows) (Rapid Golgi method, ×300) (reproduced with permission from J Neuropathol Exp Neurol)

4. Cytoskeletal changes

The neuronal cytoskeleton is composed of microfilaments, neurofilaments and microtubules which show distinctive ultrastructural characteristics (Cruz-Sánchez, 1994). Microfilaments are composed of polymers of actin with associated actin-regulatory proteins. Microtubules are important in mediating bi-directional transport of organelles and play a major role in the transport of new components in axons and dendrites (Burgoyne and Cambray-Deakin, 1988). Neurofilaments are predominantly neuronal structural elements which are modified during their lifetime by a succession of protein kinase (Nixon, 1993). The apparent greater plasticity of the neurofilament network in regions such as the perikaryon, initial segments and nodes along the axon, may provide some insight into the vulnerability of this region to neurofibrillary diseases (Nixon, 1993). Immunohistochemical studies demonstrated that a large number of antibodies, most of them monoclonal, are specific for neuronal elements in the CNS (Cruz-Sánchez, 1994).

An 8.6 KD polypeptide called ubiquitin has been linked to different intracellular stress responses most of them related to cytoskeletal abnormalities. Its covalent bond to various target proteins within the cell represents a regulatory process (Haas and Bright, 1985). Ubiquitin is implicated in the non-lysosomal degradation of abnormal proteins and other proteolytic intracellular mechanisms (Rechsteiner, 1991). Monoclonal and polyclonal antibodies have been produced to recognize the presence of ubiquitin in various pathological conditions (Lowe, 1988; Manetto, 1988; Leigh, 1989).

The cytoskeletal compromise with aging results in the presence of markers pointing to degeneration as part of a spectrum within the asymptomatic population. The spectrum of cytoskeletal changes found in aging has lately become wider. This fact has resulted in the emergence of further categorisations and subcategorisations of the degenerative process making boundaries between aging and early stages of degenerative diseases more ill-defined. In this light, Braak and Braak (1991) have defined four stages of AD on the basis of the presence of NFT, neuropil threads and components of neuritic plaques in the CNS and their topography. Changes have first been observed in the transentorhinal region, from here the process spreads in a predictable manner across the entorhinal cortex, parts of hippocampal and neocortex. Stages I and II are defined as pre-clinical, and degenerative changes are focal.

According to Vickers and co-workers (1984) age-related changes involving neurofilaments (NFs) occurring in CA1–CA2 neurons are presumably related to neurofilamentous alterations and NFT formation that have been described in subpopulations of parahippocampal neurons in the same individuals. The processes leading to neuronal degeneration in AD may therefore act upon a predisposing factor present in elderly brains. There are further progressive alterations in the normal cytoskeleton of the vulnerable subsets of neurons in the hippocampal formation and neocortex, finally resulting in the transformation of some cytoskeletal proteins into the abnormal proteins that constitute the NFT.

Scattered neocortical senile plaques have also been found in the elderly non-demented population (Arriagada et al., 1992). Yasuhara et al. (1994) described two types of dystrophic neurites (DNs) of different morphological features in SPs. Type 1 has an elongated shape, and is stained by antibodies to A68 protein, different types of phosphorylated tau, N-terminal epitopes of amyloid precursor protein (APP), and an antibody that recognizes ubiquitin. Type 2 DNs are globular in shape and stain with antibodies to chromogranin A and C-terminal epitopes of APP, as well as antibodies recognizing ubiquitin. The neocortex of elderly normal cases, have few SPs where only type 2 DNs were observed, suggesting a possible benign process.

The Lewy body is recognized as the hallmark for PD, but has also been described in 12% of asymptomatic individuals in their seventh decade, rising to 17% in the eight decade (Daniel, 1995). The finding of occasional Lewy bodies in the substantia nigra in an otherwise asymptomatic individual may be an indication of the presence of PD at a pre-clinical stage (Gibb and Lees, 1991).

In order to detect possible cytoskeletal abnormalities with aging, significant brain areas from 40 control cases aged from 17 to 101, which were selected according to previous criteria (see section 1), were immunohistologically studied with antibodies against ubiquitin, tau, and 150 and 210 phosphorylated neurofilaments. Cases were grouped according to age: group A (from 17 to 59), and group B (from 60 to 101). Areas examined were: cingulate gyrus (A23,A24) precentral cortex (A4), temporal cortex (A22), occipital cortex (A17,A18), entorhinal cortex (A26), hippocampus, caudate nucleus, lenticulate nucleus, thalamus, substantia nigra, nucleus coeruleus and medulla oblongata.

Changes related to cytoskeletal components were found in group B (Fig. 6A,B). The most common finding in this group was the presence of neurofibrillary tangles (15% of cases) which immunostained against ubiquitin, tau, and both neurofilaments subunits, followed by DNs in 10% of cases, and SPs in 5% of cases expressing ubiquitin and tau. Granulovacuolar degeneration as well as neuropil threads and Lewy bodies were not found. According to these results two cases fulfilled the diagnostic criteria for stages I–II and four cases for stages II–III of AD (Braak and Braak, 1991).

Substantia nigra did not show cytoskleletal abnormalities affecting neuronal bodies or in the neuropil.

According to these results we suggest that the cytoskeleton is not compromised during the aging process and that cytoskleletal abnormalities should correspond to a degenerative process, at least, in the nigrostriatal pathway.

5. Growth factors

Since Levy-Montalcini described nerve growth factors (NGF) in 1962, a myriad of papers have reflected the increasing interest in neurotrophins. These molecules include 4 basic proteins of low molecular weight: NGF,

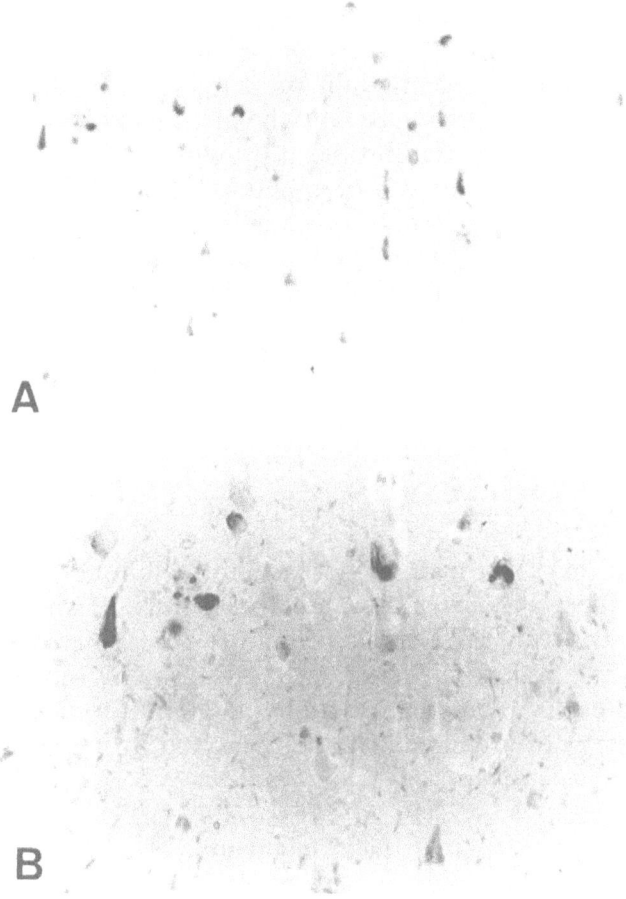

Fig. 6. Overview of entorhinal cortex showing tau deposits in the neuropil and in pyramidal neurons (Tau ×100) in a non-neuropsychiatric man aged 82 (**A**). Detail at high magnification (Tau ×250) of the same case (**B**)

brain derived neurotrophic factor (BDNF), neurotrophin 3 (NT-3) and neurotrophin 4/5 (NT-4/5) (Lindsay, 1994). In addition, there are other growth factors with important neuroprotective functions, e.g., ciliary neurotrophic factor (CNTF) and glial derived neurotrophic factor (GDNF). Until now, most studies (Jelsma and Aguayo, 1994) have been experimental and have entailed the study of the developing and damaged nervous system. Thus there is little evidence for a derangement of growth factors in aging.

Recently Rylett and Williams (1994) pointed out the age-related, region-specific decrement of NGF effects upon cholinergic neurons in the rat which may be extrapolated to cognitive impairment in the aging human. These authors highlighted the limitation for therapy that this decrement in responsiveness could lead to.

The rat nigrostriatal system is responsive to BDNF, NT-3, NT-4/5 and GDNF. Neurons are "rescued" from death after injury by infusion of growth factors (Lindsay, 1995; Tomac et al., 1995; Beck et al., 1995; Cheng and Mattson, 1994). Unfortunately, all studies were made under pathological conditions (axotomy, infusion of toxic substances, etc.) and the possible effects of these substances in aging humans remains to be demonstrated.

Williams et al. (1994) demonstrated the effectiveness of N- Cell adhesion molecule (N-CAM) and basic fibroblast growth factor (bFGF) in promoting sprouting of dendrites however a therapeutic role (bFGF and N-CAM) and influence on aging still remain to be elucidated. Linnemann et al. (1993) found no changes in levels of NCAM mRNA in rats from the 40th day postnatal to old age. With the Golgi method, we demonstrated alterations in the dendritic network of the nigrostriatal pathway in aging (Cruz-Sánchez et al., 1995). The implications of these changes, and the possible relationship between them, need to be further elucidated.

6. Glial changes

In the aging rat there is a relative increase in the number of astrocytes in all brain regions. These astrocytes show reactive features (Landfield et al., 1994). The amount of glial fibrillary acidic protein (GFAP) increases with aging in rat brain regions, reflecting generalized gliosis. On the other hand, the gliotrophic protein S-100 increases from young adult to maturity in all brain regions and decreases in old age (Linnemann and Skarsfelt, 1994).

7. Neuropigments

The accumulation of lipofuscin in neuronal cells during aging is one of the most constant findings in aging. Some authors have related the accumulation of lipofuscin to the antioxidant/oxidant ratio. The accumulation of lipofuscin is less when there is an induction of brain glutathione reductase. It seems that the maintenance of a high antioxidant/oxidant ratio is essential to accumulation of lipofuscin (Lopez-Torres et al., 1993). Interestingly, this theory joins the free radical theory of aging with the accepted standard manifestation of aging i.e. lipofuscin accumulation.

Neuromelanin decreases after middle age. This decrease is thought to be due to selective atrophy of the most pigmented cells rather than to a generalized loss of pigments (Mann and Yates, 1974).

8. Plasticity of the nervous system

Aging of the nervous system is characterized by reduced neuronal plasticity. For example, Crutcher (1990) stated that the age-related decrease in sympathetic sprouting is primarily due to decreased target receptivity. When transplanted into a young host, spinal ganglion cells survive longer whether the donor is young or old.

9. Molecular changes

In a seminal manuscript published in Scientific American, Selkoe (1992) states of the molecular theory of aging: "cells throughout the body senesce because defects slowly accrue in their DNA". An increasing number of papers are in agreement with this theory.

Neurotransmission changes in age have only been known in the last few years. Cyclic AMP- dependent protein kinases and protein kinase C, the final pathway of many neurotransmitters, have been reported to be altered in aging.

The role of phosphorylation in the modification of protein function is being increasingly studied (Magnoni et al., 1991). The metabolism of the very genetic material itself is affected. Depression of normally silent genes, decreased synthesis of neuronal mRNA, altered post-transcriptional splicing of mRNA, reduced turnover and accumulation of non-functional proteins are some of the possible aging mechanisms (Finch and Morgan, 1990). During aging there is also induction of DNA transcription factors. AP-1 transcription factor DNA binding activity is strongly induced during aging in the presence of anoxia. According to Kaminska and Kaczmarek (1993) the AP-1 transcription factor is a dimer of c-Fos and c-Jun oncogenes products; these authors shows that not all the molecular machinery becomes defective during aging. Our task is to decide which are the crucial alterations of aging and which represent epiphenomena or compensatory mechanisms.

The alterations in cell membrane composition in normal aging seem to result in impaired neurotransmitter-triggered signal transduction. The impaired signal transduction seems to be related to dysfunctions in the coupling of G proteins to their receptors and effectors. In AD, cell membranes show a 30% reduction in cholesterol/phospholipid composition (Roth et al., 1995).

There are other processes involved in aging: enhanced glycation of proteins, increased production of oxygen-derived radicals, decreased ability to defend against the accumulation of free radicals (Smith et al., 1995).

Final comments

In the present review, we have focused on only one aspect of the aging syndrome – the aging of the nigrostriatal pathway. This shares some features with the aging of other systems but probably has its own specific features derived from its peculiar molecular characteristics and/or from a selective vulnerability to aging.

The association between the aging syndrome and a pathological process is likely to result in the compounding of signs and symptoms which is far more complex than initially thought.

To gain further insight into the aging syndrome, the acquisition of rigorous criteria for selecting control cases is paramount. It is necessary to exclude pathological cases, patients with a prolonged agonal state and patients with systemic diseases which may affect the CNS. Post-mortem delay has to be as short as possible, thus implying the closest collaboration between clinicians and pathologists. Improving methods of preservation of tissues is also crucial.

The molecular side of the work began a few years ago and a wide variety of results have ensued (lipofuscin accumulation, neuronal loss and atrophy, glia proliferation, synaptic changes and so on), thus, definitive conclusion are lacking. Now, molecular biologists are beginning to understand the real causes of the aging process. Free radicals and oxidative stress, growth factors, DNA or RNA damage are crucial to the aging process. The next few years may lead to previously unexpected pathophysiological and therapeutic discoveries. Thus, it will hopefully be possible to provide a better quality of life to aging people and prolong life itself. As Selkoe (1992) says, the aim of neuroscientists is to improve the quality of life of elderly people.

Perhaps the complexity of the aging syndrome will go beyond the neuroscientific field of study and will involve the work of other investigators. Muscle aging, for example, could influence directly the aging of the nervous system and similarly could influence cardiovascular aging. Thus, researchers should be more open to collaboration with scientists in other fields.

We must keep in mind that aging is the process that causes more suffering to people, much more than any disease, and thus needs to be studied to a greater extent, as our aim is to alleviate suffering.

Acknowledgement

This paper was supported by "Fondo de Investigacion Sanitaria", Spanish health ministry (FIS 96/1320).

References

Arriagada PV, Marzloff K, Hyman BT (1992) Distribution of Alzheimer-type pathologic changes in nondemented elderly individuals matches the pattern in Alzheimer's disease. Neurology 42: 1681–1688

Beck KD, Valverde J, Alexi T, Poulsen K, Moffat B, Vandlen R, Rosenthal A, Hefti F (1995) Mesencephalic dopaminergic neurons protected by GDNF from axotomy-induced degeneration in the adult brain. Nature 373: 339–341

Boatell MLl, Mahy N, Cardozo A, Ambrosio S, Tolosa E, Cruz-Sánchez FF (1992) Neuronal changes in the nigrostriatal pathway of 1-Methl-4-Phenylpyridine- treated mice. Meth Find Exp Clin Pharmacol 14(10): 781–787

Braak H, Braak E (1991) Neuropathological stageing of Alzheimer-related changes. Acta Neuropathol 82: 239–259

Burgoyne RD, Cambray-Deakin MA (1988) The cellular neurobiology of neuronal development: the cerebellar granule cell. Brain Res Rev 13: 77–101

Calne DB, Peppard RF (1987) Aging of the nigrostriatal pathway in humans. Can J Neurol Sci 14: 424–427

Calne DB, Eisen A, McGeer E, Spencer P (1986) Alzheimer's disease, Parkinson's disease and motor neurone disease: abiotrophic interaction between ageing and environment? Lancet (Nov 8): 1067–1070

Cheng B, Mattson MP (1994) NT-3 and BDNF protect CNS neurons against metabolic/excitotoxic insults. Brain Res 640: 56–67

Crutcher KA (1990) Age-related decrease in sympathetic sprouting is primarily due to decreased target receptivity: implications for understanding brain aging. Neurobiol Aging 11: 175–183

Cruz-Sánchez FF (1994) Antigenic determinant properties of neurofibrillary tangles. Relevance to progressive supranuclear palsy. J Neural Transm [Suppl] 42: 165–178

Cruz-Sánchez FF, Cardozo A, Ambrosio S, Tolosa E, Mahy N (1993) Plasticity of the nigrostriatal system in MPTP-treated mice. Mol Chem Neuropathol 19: 163–176

Cruz-Sánchez FF, Cardozo A, Tolosa E (1995) Neuronal changes in the substantia nigra with aging: a Golgi study. J Neuropathol Exp Neurol 54: 74–81

Daniel SE (1995) Parkinson's disease. In: Cruz-Sánchez FF, Ravid R, Cuzner ML (eds) Neuropathological diagnostic criteria for brain banking. IOS Press, Amsterdam

Fearnley JM, Lees AJ (1991) Ageing and Parkinson's disease: substantia nigra regional selectivity. Brain 114: 2283–2301

Finch CE, Morgan DG (1990) RNA and protein metabolism in the aging brain. Ann Rev Neurosci 13: 75–87

Gibb WRG, Lees AJ (1991) Anatomy, pigmentation, ventral and dorsal subpopulations of the substantia nigra, and differential cell death in Parkinson's disease. J Neurol Neurosurg Psychiatry 54: 388–396

Gsell W, Lange KW, Pfeuffer R, Heckers S, Heisen H, Senitz D, Jellinger K, Ransmayr G, Wichart I, Vock R, Beckmann H, Riederer P (1993) How to run a brain bank. A report from the Austro-German brain bank. J Neural Transm [Suppl] 39: 31–70

Haas AL, Bright PM (1985) The immunochemical detection and quantification of intracellular ubiquitin-protein conjugates. J Biol Biochem 260: 12464–12473

Hodge CF (1894) Changes in the ganglion cells from birth to senile death. Observations on man and honey-bee. J Physiol 17: 129–134

Jelsma TN, Aguayo AJ (1994) Trophic factors. Curr Opin Neurobiol 4: 717–725

Kaminska B, Kaczmarek L (1993) Robust induction of AP-1 transcription factor DNA binding activity in the hippocampus of aged rats. Neurosci Lett 153: 189–191

Landfield PW, Rose G, Sandles L, Wohlstadter TC, Lynch G (1994) Patterns of astroglial hypertrophy and neuronal degeneration in the hippocampus of aged, memory-deficient rats. J Gerontol 32: 3–12

Leigh PN, Probst A, Dale GE, Power DP, Brion J-P, Dodson A, Anderton BH (1989) New aspects of the pathology of neurodegenerative disorders as revealed by ubiquitin antibodies. Acta Neuropathol (Berl) 79: 61–72

Lindsay RM (1995) Neuron saving schemes. Nature 373: 289–290

Lindsay RM, Wiegand SJ, Altar A, DiStefano P (1994) Neurotrophic factors: from molecule to man. TINS 17: 182–190

Linnemann D, Skarsfelt T (1994) Regional changes in expression of NCAM, GFAP, S100 in aging rat brain. Neurobiol Aging 15: 651–655

Linnemann D, Gaardsvoll H, Olsen M, Bock E (1993) Expression of NCAM mRNA and polypeptides in aging rat brain. Int J Dev Neurosci 11: 71–81

Lopez Torres M, Perz Campo R, Fernandez A, Barba C, Barja de Quiroga G (1993) Brain glutathione reductase induction increases early survival and decreases lipofuscin accumulation in aging frogs. J Neurosci Res 34: 233–242

Lowe J, Blanchard A, Morrel K, Lennox G, Reynolds L, Billet M, Landon M, Mayer RJ (1988) Ubiquitin is a common factor in intermediate filament inclusion bodies of diverse type in man, including those of Parkinson's disease, Pick's disease and Alzheimer's disease, as well as Rosenthal fibres in cerebellar astrocytomas, cytoplasmic bodies in muscle and Mallory bodies in alcoholic liver disease. J Pathol 155: 9–15

Machado-Salas J, Scheibel ME, Scheibel AB (1977) Neuronal changes in the aging mouse: spinal cord and lower brain stem. Exp Neurol 54: 504–512

Magnoni MS, Govoni S, Battaini F, Trabucchi M (1991) The aging brain: protein phosphorylation as a target of changes in neuronal function. Life Sci 48: 373–385

24 F. F. Cruz-Sánchez et al.

Mann DMA, Yates PO (1974) Lipoprotein pigments – their relationship to ageing in the human nervous system. II. The melanin content of pigmented nerve cells. Brain 97: 489–498

Mann DMA, Yates PO, Marcyniuk B (1984) Monoaminergic neurotransmitter systems in presenile Alzheimer's disease and in senile dementia of Alzheimer type. Clin Neuropathol 3(5): 199–205

Mannetto V, Perry G, Tabaton M, Mulvihill P, Fried VA, Smith HT, Gambetti P, Autulio Gambetti L (1988) Ubiquitin is associated with abnormal cytoplasmic filaments characteristic of neurodegenerative diseases. Proc Natl Acad Sci USA 85: 4501–4505

McGeer PL, McGeer EG, Suzuki JS (1977) Aging and the extrapyramidal function. Arch Neurol 34: 33–35

Migheli A, Attanasio A, Pezzulo T, Gullotta F, Giordana MT, Schiffer D (1992) Age related ubiquitin deposits in dystrophic neurites: an immunoelectron microscopic study. Neuropathol Appl Neurobiol 18: 3–11

Moral A, Cardozo A, Rossi M, Tolosa E, Cruz-Sánchez FF (1994) Spinal motor neurons in aging: a morphological and immunohistochemical study (Abstract). Brain Neuropathol 4: 577

Nixon RA (1993) The regulation of neurofilament protein dynamics by phosphorylation: clue to neurofibrillary pathobiology. Brain Pathol 3: 29–38

Patt S, Gertz HJ, Lieselote G, Cervós-Navarro J (1991) Pathological changes in dendrites of substantia nigra neurons in Parkinson's disease: a Golgi study. Histol Histopathol 6: 373–380

Ramón-Moliner E (1970) The Golgi-Cox technique. In: Nauta WJH, Ebbesson SOE (eds) Contemporary research methods in neuroanatomy. Springer, New York, pp 32–55

Ravid R, Swaab DF, van Zwieten EJ, Salehi A (1995) Controls are what makes a brain bank go round. In: Cruz-Sánchez FF, Ravid R, Cuzner ML (eds) Neuropathological diagnostic criteria for brain banks. IOS Press, Amsterdam, pp 4–13

Rechsteiner M (1991) Natural substrate of the ubiquitin proteolytic pathway. Cell 66: 615–618

Roth GS, Joseph JA, Preston Mason R (1995) Membrane alterations as causes of impaired signal transduction in Alzheimer's disease and aging. TINS 18: 203–206

Rylett RJ, Williams LR (1994) Role of neurotrophins in cholinergic-neurone function in the adult and aged CNS. TINS 17: 486–490

Selkoe DJ (1992) Aging brain, aging mind. Sci Am 267: 134–142

Smith MA, Sayre LM, Monnier VM, Perry G (1995) Radical ageing in Alzheimer's disease. TINS 18: 172–176

Scheibel ME, Scheibel AB (1970) The rapid Golgi method. Indian summer or renaissance? In: Nauta WJH, Ebbesson SOE (eds) Contemporary research methods in neuroanatomy. Springer, New York, pp 1–11

Scheibel ME, Lindsay RD, Tomiyasu U, Scheibel AB (1976) Progressive dendritic changes in the aging human limbic system. Exp Neurol 47: 392–403

Scheibel ME, Lindsay RD, Tomiyasu U, Scheibel AB (1977) The aging human Betz cell. Exp Neurol 56: 598–609

Teräväinen H, Calne DB (1983) Motor system in normal aging and Parkinson's disease. In: Katzman R, Terry R (eds) The neurology of aging. Davis, Philadelphia, pp 85–109

Thiessen B, Rajput AH, Laverty W, Desai H (1990) Age, environments and the number of substantia nigra neurons. In: Streifler MB, Korczyn AD, Melamed E, Youdim MBH (eds) Parkinson's disease: anatomy, pathology, and therapy. Raven Press, New York, pp 201–206 (Adv Neurol 53)

Tomac A, Lindqvist E, Lin LF H, Ögren SO, Young D, Hoffer BJ, Olson L (1995) Protection and repair of the nigrostriatal dopaminergic system by GDNF in vivo. Nature 373: 335–339

Tomlinson BE, Irving D (1977) The number of motor neurons in the human lumbosacral cord throughout life. J Neurol Sci 34: 213–219

Valverde F (1970) The Golgi method. A tool for comparative structural analyses. In: Nauta WJH, Ebbesson SOE (eds) Contemporary research methods in neuro-anatomy. Springer, New York, pp 12–31

Vickers JC, Riederer BM, Marugg RA, Buée-Scerrer V, Buée L, Delacourte A, Morrison JH (1994) Alterations in neurofilament protein immunoreactivity in human hippocampal neurons related to normal aging and Alzheimer's disease. Neurosci 62(1): 1–13

Wiedenmann B, Franke W (1985) Identification and location of synapthophysin, an integral membrane glycoprotein of Mr 38,000 characteristic of presynaptic vesicles. Cell 41: 1017–1028

Williams EJ, Furness J, Walsh FS, Doherty P (1994) Activation of FGF receptor underlies neurite outgrowth stimulated by L1, N-CAM and N-cadherin. Neuron 13: 583–594

Yasuhara O, Kawamata T, Yoshinari A, McGeer EG, McGeer PL (1994) Two types of dystrophic neurites in senile plaques of Alzheimer's disease and elderly non-demented cases. Neurosci Lett 171: 73–76

Authors' address: F. F. Cruz-Sánchez, M.D., Banco de Tejidos Neurológicos, Service of Neurology, Hospital Clinico y Provincial, Villarroel 170, Barcelona 08036, Spain

Psychiatric symptoms and behavioural disturbances in the dementias

A. Burns

Department of Psychiatry, University of Manchester, Withington Hospital,
Manchester, United Kingdom

Summary. This contribution will summarise the relationship between psychiatric symptoms and behavioural disturbances in dementia. The different types of symptoms will be discussed with particular relation to Alzheimer's and Parkinson's disease and results will be presented in relation to clinical symptomatology in the Lewy body variant of Alzheimer's disease.

Introduction

There is increasing appreciation that the symptoms and signs of dementia are myriad and consist of features attributable directly to defects in cognitive function such as amnesia, aphasia, apraxia and the agnosias, as well as a variety of symptoms apparently unrelated to such symptoms. These so-called "non-cognitive" features (Burns et al., 1990) can be divided into two main categories. *Behaviours* are observed by others (examples are aggression, wandering, and eating disorders), whereas *psychiatric symptoms* (psychopathology, phenomenology) are inner experiences unique to the subject (examples are hallucinations, delusions, and affective disturbances). Clearly, there can be significant overlap, for example, responding to an auditory hallucination may cause a patient to become aggressive or an agitated depression may be manifest as wandering or pacing.

There are a number of different types of symptom, summarised in Table 1, which follows the classification outlined by Cummings and Victoroff (1990).

Description of symptoms

Delusions

These can be of a number of types and range from simple ideas of theft of personal possessions which can be an understandable consequence of a memory disorder (i.e., misplacing a handbag and forgetting where it is, leading to a belief it has been stolen, often culminating in an accusation of a

Table 1. Non-cognitive features of dementia

— Delusions
— Hallucinations
— Misidentifications
— Reduplications
— Mood disorders
— Neurovegetative changes
— Psychomotor changes
— Personality changes

particular individual) or can be a complex systematised delusional elaboration combining hallucinations and delusional ideation. Specific delusional ideas include the belief that someone famous is in love with the person (de Clérambault's syndrome), that loved ones have been replaced by impostors (Capgras syndrome), that one's house is not one's own home, the belief that the subject has been abandoned (sadly, this occasionally has a basis in truth), infestation by animals (parasitosis), that another person has taken up residence in the house (Phantom boarder syndrome) and the belief that a spouse is having an affair. Delusions occur in up to 30% of patients with Alzheimer's disease. Some studies have found that delusions of theft are more common in men. Systematised (i.e., complex) delusions seem to require a relatively intact cerebral cortex (Jacoby and Levy, 1980) and have been associated with basal ganglia calcification (Burns et al., 1990). Some studies have found that delusions, when in conjunction with hallucinations, predict more rapid cognitive decline (Stern et al., 1994).

Hallucinations

The most common modalities are auditory and visual hallucinations although olfactory and haptic perceptions have also been described. Visual hallucinations can occur as part of a confusional state but can also be present as part of the dementia syndrome. Auditory hallucinations can sometimes give commands to the subject resulting in violence. Patients with hallucinations show a more rapid deterioration in cognitive function over time (Burns et al., 1990).

Misidentification

These are of four main types: misidentification of other people; the belief that events occurring on the television are in real three-dimensional space; the belief that others are in the house (this could as easily be classified as a delusion); and the mirror sign where subjects converse with their own mirror image or can occasionally become frightened and aggressive when seeing their own reflection, supposing it is another individual meaning them harm. These have only been described in a few studies but prevalence rates of up to 30 per cent have been documented. Misidentifications have been associated with younger age of onset of disease but longer survival (Burns et al., 1991).

Reduplication

Patients can sometimes complain that things have been duplicated. Often these are relatives or friends but can be inanimate objects such as houses, clothes or even pets.

Mood disorders

There have been a number of studies looking at the prevalence of depression in dementia (Allen and Burns, 1995). Wide discrepancies in the rates of depression have been described which are not completely accounted for by differences in sample selection or methods of case ascertainment (whether it be for depression or dementia). Generally, mood disorders are common but this in part may be due to the similarity of symptoms in the two disorders. Indeed, in DSM III R, diagnoses of major depression and primary degenerative dementia of the Alzheimer-type were mutually exclusive (Burns, 1990). Rates of up to 86 per cent have been found and usually these are higher in the milder stages of the disease, often mistakenly assumed to be due to the retention of insight with the inevitable consequences on mood and morale. Irritability, aggression and apathy can co-exist in patients with Alzheimer's disease and Huntington's disease (Burns et al., 1990b). Patients with depression and Alzheimer's disease have less severe atrophy on computed tomography scans and have different neuropathological profiles — relatively less severe neuronal fall-out in the locus coeruleus and relative preservation of neurones in the nucleus basalis of Meynert (Förstl et al., 1992, vide infra).

Patients with symptoms of mania are rare in comparison (only 2–3%) and may have atrophy of the frontal regions on computed tomography scans.

Neurovegetative and psychomotor changes

These encompass a number of different behaviours which tend to accompany the later stages of dementia. Neurovegetative functions include features such as disturbances of sleep, appetite and sexual function. Psychomotor changes consist of pacing and wandering, agitation, purposeless hand movements and various types of disturbed motor activity. The Klüver-Bucy syndrome describes a disorder originally found in monkeys who had bi-temporal lobotomy and consists of a cluster of symptoms — sexual disinhibition, visual agnosia, hyperorality, withdrawal/apathy, "rage" behaviour, hypermetamorphosis, and binge eating. The syndrome has been reported in both Alzheimer's disease and Pick's disease (Cummings and Duchen, 1991).

Personality changes

These have probably been the least researched of all the non-cognitive features in dementia but are of important as they may be one of the earliest changes. The original Blessed scale included personality changes (Blessed et al., 1968) as part of its measure of dementia and more recently, a scale (Table 2) assessing personality changes in patients post head injury has shown alterations in patients with dementia (Petry et al., 1988, 1989).

Importance of non-cognitive features

Non-cognitive features in dementia are of importance for a number of reasons. They can be of help in the *differential diagnosis* of dementia, with one recent example of this being the proposed diagnostic criteria for dementia of the frontal lobe type where a number of non-cognitive features were cited as being integral to the clinical diagnosis of the syndrome (Brun et al., 1994). These are summarised in Table 3.

The role of non-cognitive features in *early diagnosis* has yet to be fully exploited, partly due to the predominance of cognitive symptoms in diagnostic systems for dementia. It is likely that subtle changes in personality predate cognitive changes. Oppenheim (1994) found in a third of cases of Alzheimer's disease, symptoms of a psychiatric type (e.g. withdrawn demeanour, paranoid ideas, anxiety, irritability and aggression).

Subtypes of Alzheimer's disease can be shown by clinico-pathological studies. Förstl et al. (1992, 1993, 1994) in a series of studies reported clinico-pathological correlations in a prospectively examined group of over 50 pa-

Table 2

	−2 −1 0 +1 +2	
Out-of-touch	I—I—I—I—I	Down-to-earth
Relies on other	I—I—I—I—I	Does things himself
Childish	I—I—I—I—I	Mature
Listless	I—I—I—I—I	Enthusiastic
Changeable	I—I—I—I—I	Stable
Unreasonable	I—I—I—I—I	Reasonable
Lifeless	I—I—I—I—I	Energetic
Unhappy	I—I—I—I—I	Happy
Cold	I—I—I—I—I	Affectionate
Cruel	I—I—I—I—I	Kind
Irritable	I—I—I—I—I	Easygoing
Mean	I—I—I—I—I	Generous
Dislikes company	I—I—I—I—I	Fond of company
Quiet	I—I—I—I—I	Talkative
Insensitive	I—I—I—I—I	Sensitive
Quick-tempered	I—I—I—I—I	Even-tempered
Excitable	I—I—I—I—I	Calm
Rash	I—I—I—I—I	Cautious

Table 3. Non-cognitive features in fronto-temporal dementia
(from Brun et al., 1994)

Behaviour disorder
— Disinhibition
— Mental rigidity/inflexibility
— Hyperorality
— Stereotyped behaviour
— Utilisation behaviour
— Distractibility
Affective symptoms
— Depression/sentimentality
— Delusion
— Hypochondriasis
— Emotional unconcern
— Amimia

Table 4 (adapted from Förstl et al., 1992)

	Alzheimer's disease	
	Depressed (N = 14)	Non-depressed (N = 38)
Age onset	77.1	75.0
Duration (years)	8.2	7.5
MMSE (Max = 30)	7.4	4.2*
CAMCOG (Max = 107)	24.6	12.6*
Neuronal counts		
Locus coeruleus	36.9	51.4*
Nucleus basalis of Meynert	95.1	71.2
Substantia nigra	495.5	478.1

*Significant difference at $p < 0.05$. All others non
significant

tients suffering from autopsy-proven Alzheimer's disease. The results in relation to depression are summarised in Table 4. Depression was found in 14 out
of 52 patients and was associated with lower neuronal counts in the locus
coeruleus and a trend to higher counts in the nucleus basalis of Meynert. This
is of theoretical importance, as a disproportionate noradrenergic deficit, as
evidenced by neuronal fall-out in the locus coeruleus, would predispose
to depression. Although patients with depression were significantly less
cognitively impaired than the non-depressed group, depression contributed
significantly to the variance of neuronal counts in the locus coeruleus, even
when co-varying for gender, age of onset, cognitive impairment and cortical
Alzheimer plaques.

Behavioural disturbances have also been found to have biological correlates. Fifty-four patients were examined in this part of the study and associa-
tions were sought between clinical ratings on the Stockton Geriatric Rating

Scale (SGRS) and pathological measures. This scale has four factors which have been validated on an elderly hospitalised population with dementia, — physical disability (P), apathy (A), communication failure (C), and socially irritating behaviour. The first three factors have been shown to be associated with poor survival and are associated with more severe dementia. A combined score on the PAC factors is associated with lower brain-weight, more severe tangle pathology in the parahippocampal gyrus and the frontal and parietal cortex, and lower neuronal counts in the hippocampus and basal nucleus of Meynert. Two features of the Klüver-Bucy syndrome (rage behaviour and hypermetamorphosis) were significantly associated with lower counts of large neurons in the parahippocampal gyrus and parietal neocortex but not with pathological changes in the subcortical nuclei. The findings suggested that decreased cortical inhibition of relatively preserved subcortical structures may be the basis of some of the features of the Klüver-Bucy syndrome. Generally, the correlation between ratings of behavioural disturbance accompanying severe dementia and measures of pathological change confirms the association between the severity of the clinical syndrome and the pathological changes.

Psychotic phenomena has been related to neuropathological changes. Of 56 patients, 13 had hallucinations, paranoid delusions in 9, and delusional misidentification in 14. Misidentifications were associated with lower neuronal counts in the CA1 area of the hippocampus. Delusions and hallucinations were seen in patients with less severe cell loss in the parahippocampal gyrus and the dorsal raphé nucleus. Basal ganglia calcification was particularly common in patients with delusions and delusional misidentifications. Lower neuronal counts in the pyramidal cell layer of the CA1 region accords well with the hypothesis of delusional misidentification being a disconnection of memories. The trend for patients with hallucinations to have lower neuronal counts in the dorsal raphé nucleus is in keeping with neurochemical studies showing lowered serotonin concentrations in the brain. Such neurochemical and neuropathological changes may provide a model by which the functional psychoses can be better understood.

One of the main reasons that non-cognitive features are of relevance in dementia is that they cause often unacceptable *strain on carers*. There is a large literature to show that non-cognitive features place disproportionate stress on carers and it is particularly behavioural disturbances which do so. Incontinence, wandering and nocturnal disturbance place the greatest stress on carers (Sanford, 1975).

Finally, while there is no effective treatment for the cognitive disorder which accompanies Alzheimer's disease, there are *effective treatments* for non-cognitive features (Burns, 1993).

Parkinson's disease

It has long been known that patients with Parkinson's disease experience a number of psychiatric features. Depression is the most widely quoted, up to 40 per cent of patients being affected, possibly due to an alteration in cerebral serotonin production (Mayeux et al., 1986). Psychotic symptoms have been regarded as much less common and are usually caused by anti-parkinsonian medication (Baldwin and Byrne, 1989). However, a recent review of the literature suggests that psychosis may be commoner than previously thought and the role of drugs needs to be more fully evaluated (Burns and Like, in preparation). For example, in a pre-levodopa study, Monroe et al. (1951) found that 33 per cent of patients presenting with Parkinson's disease were "psychotic". Damasio et al. (1971) found that 15 out of 48 patients were psychiatrically disturbed, but 13 of the 15 had a previous history of psychiatric illness. A review of some of the published literature found visual hallucinations in 19 per cent of patients (10 studies with a total of 1,799 patients) and delusions in 11 per cent (6 studies including 344 patients).

Dementia of Lewy body type

Lewy body disease has been well documented and clinical descriptions will be found elsewhere in this symposium. Psychiatric symptoms occur particularly commonly in this type of dementia (McKeith et al., 1992), in particular visual hallucinations and delusions. Frederich H. Lewy (1885–1950) described cases of paralysis agitans in three publications (1912 – 25 cases, 1913 – 60 cases, and 1923 – 88 cases), and his full clinical and pathological summary of 43 patients (on whom all information was available) contains some information as to the proportion of cases of Parkinson's disease who had psychiatric disease.

Of the 43 patients, 21 were regarded as demented, 10 suffered an affective disorder of whom three were paranoid, three had hallucinations and two had

Table 5. The Lewy body variant of Alzheimer's disease (from Förstl et al., 1993)

Neuropsychological	LB+ (N = 8)	LB− (N = 8)
Neuropsychological		
MMSE (Max = 30)	9.8	7.5
CAMCOG (Max = 107)	26.6	23.0
Computed tomography		
Frontal s/d space (mls)	34.7	18.4*
Lateral ventricles (mls)	123	135
Neuropathology		
Brain weight (g):		
Hemispheres	1,274	1,182
Brain stem	165	162
Tangles/sq mm:		
P/H gyrus	2.3	13.1**
Dentate gyrus	2.6	9.4
Frontal lobe	2.5	11.4*

*Significant difference at p < 0.03; **significant difference at p < 0.003. All others non significant. s/d subdutal

mild cognitive impairment. Only 12 were deemed to be psychiatrically "normal" (Förstl and Lewy, 1991). Of the demented group, the majority had neocortical Alzheimer pathology. Only one patient had Lewy bodies in the neocortex.

Förstl et al. (1993) reported on eight patients with Lewy body dementia and Alzheimer's disease and compared them to eight age- and sex-matched patients with Alzheimer's disease alone. The patients with additional Lewy bodies had the same degree of cognitive impairment but had more frontal lobe atrophy on CT scan and less tangle pathology in the parahippocampal gyrus and frontal lobe. The results are summarised in Table 5.

Conclusions

In conclusion, psychiatric symptoms and behavioural disturbances are a common accompaniment to dementia with up to one-third of patients being affected at some point during their illness. Biological correlates have been found within the dementias and in Alzheimer's disease in particular which emphasise the fact that they are not merely epiphenomena. There is a lack of structured and standardised instruments to measure these non-cognitive features and more quantified and accurate descriptions are needed. Finally, these symptoms are important as they cause stress to carers and are a major determinant of institutionalisation.

References

Allen H, Burns A (1995) The non-cognitive features of dementia. Rev Clin Gerontol 5: 57–75

Baldwin R, Byrne J (1989) Psychiatric aspects of Parkinson's disease. Br Med J 299: 3–4

Blessed G, Tomlinson B, Roth M (1968) The association between quantitative measures of dementia and of senile change in the cerebral grey matter of elderly subjects. Br J Psychiatry 114: 797–811

Burns A (1990) Disorders of affect in Alzheimer's disease (Editorial). Int J Geriatr Psychiatry 5: 63–66

Burns A, Jacoby R, Levy R (1990) Psychiatric phenomena in Alzheimer's disease. Br J Psychiatry 257: 72–94

Burns A, Folstein S, Brandt J, Folstein M (1990) Clinical assessment of irritability, apathy and aggression in Alzheimer's and Huntington's disease. J Nerv Ment Dis 178: 20–26

Burns A, Lewis G, Jacoby R, Levy R (1991) Survival in Alzheimer's disease. Psychol Med 21: 363–370

Brun A, Englund E, Gustafsson L, et al (1994) Clinical and neuropathological criteria for frontotemporal dementia. The Lund and Manchester Groups. J Neurol Neurosurg Psychiatry 57: 416–418

Cummings J, Duchen L (1981) Klüver-Bucy syndrome in Pick disease. Neurol 31: 1415–1422

Cummings J, Victoroff J (1990) Non-cognitive neuropsychiatric syndromes in Alzheimer's disease. J Neuropsychiatr Neuropsychol Behav Neurol 3: 140–158

Damasio A, Tunes J, Macedo C (1971) Psychiatric aspects in parkinsonism treated with L-Dopa. J Neurol Neurosurg Psychiatry 34: 502–507

Förstl H, Levy R (1991) F.H. Lewy on Lewy bodies, parkinsonism and non-dementia. Int J Geriatr Psychiatry 6: 757–766

Förstl H, Burns A, Luthert P, Cairns N, Lantos P, Levy R (1992) Clinical and neuropathological correlates of depression in Alzheimer's disease. Psychol Med 22: 877–884

Förstl H, Burns A, Luthert P, Cairns N, Levy R (1993) The Lewy Body variant of Alzheimer's disease — clinical and neuropathological findings. Br J Psychiatry 162: 385–392

Förstl H, Burns A, Levy R, Cairns N, Luthert P, Lantos P (1993) Neuropathological correlates of behavioural disturbance in confirmed Alzheimer's disease. Br J Psychiatry 163: 364–368

Jacoby R, Levy R (1980) Computed tomography in the elderly 2. Senile dementia: diagnosis and functional impairment. Br J Psychiatry 136: 256–269

Mayeux R, Stern Y, Williams J, et al (1986) Clinical and biochemical features of depression in Parkinson's disease. Am J Psychiatry 143: 757–759

McKeith I, Perry R, Fairbairn A, Jabeen S, Perry E (1992) Operational criteria for senile dementia of Lewy body type. Psychol Med 22: 911–922

Munro RT (1951) Disease in old age. Cambridge and Massachusetts

Oppenheim G (1994) The earliest signs of Alzheimer's disease. J Geriatr Psychiatry Neurol 7: 118–122

Petry S, Cummings J, Cummings L, Hill MA, Shapira J (1989) Personality alterations in dementia of the Alzheimer type: a three-year follow-up study. J Geriatr Psychiatry Neurol 2, 3: 184–188

Sanford J (1975) Tolerance of debility in elderly dependants by supports at home. Br Med J 3: 471–473

Stern Y, Albert M, Brandt J, et al (1994) Utility of extrapyramidal signs and psychosis as predictors of cognitive and functional decline — nursing home admission and death in Alzheimer's disease. Neurology 44: 2300–2307

Author's address: Prof. A. Burns, Department of Psychiatry, University of Manchester, Withington Hospital, West Didsbury, Manchester M20 8LR, United Kingdom

PET and the investigation of dementia in the parkinsonian patient

N. Turjanski and **D. J. Brooks**

MRC Cyclotron Unit, Hammersmith Hospital, London, United Kingdom

Summary. Parkinsonism and dementia are present in a number of neurodegenerative conditions. They may be a manifestation of isolated brain stem (Parkinson's disease) or diffuse Lewy body disease (DLBD), or be secondary to combined Lewy body and Alzheimer's disease (AD) pathologies. Positron emission tomography (PET) studies show a resting pattern of fronto-temporo-parietal hypometabolism in both, AD and in parkinsonism-dementia (PD-dementia) patients, even when only isolated brain stem Lewy body disease is found at pathology. We have studied three patients fulfilling clinical criteria for diagnosis of DLBD. Their ^{18}F-fluorodeoxyglucose (FDG) PET results showed an AD pattern of fronto-temporo-parietal hypometabolism, though these patients had only mild cognitive dysfunction. Parkinsonism associated with apraxia is observed in corticobasal degeneration (CBD) while impairment of frontal functions, such as planning and sorting, is seen in patients with progressive supranuclear palsy (PSP). PET studies in CBD patients have shown an asymmetric hypometabolism of cortex and thalamus contralateral to the affected limbs, while in PSP patients there is a global metabolic reduction most pronounced in frontal areas and the basal ganglia. These results suggest that metabolic PET studies can help to distinguish PD-dementia, PSP and CBD, but are unable to distinguish PD-dementia from AD. Further studies with post-mortem confirmation are required to establish if DLBD is associated with a distinctive pattern of resting hypometabolism.

Introduction

The combination of dementia and parkinsonism occurs in patients during the evolution of both Parkinson's disease (PD) and Alzheimer's disease (AD). Dementia is thought to occur in 10–20% of patients with PD (Brown and Marsden, 1984; Lees, 1985; Tison et al., 1995) while extrapyramidal signs are found in 30–90% of patients with AD (Molsa et al., 1984; Mayeux et al., 1985; Ditter and Mirra, 1987; Tyrrell et al., 1990). Post-mortem examinations of these patients have shown isolated brain stem or diffuse Lewy body disease, isolated Alzheimer's type changes, or a mixture of both pathologies. It remains unclear whether diffuse Lewy body disease (DLBD) is a distinct clinico-pathological entity or a phenotypic expression of PD.

DLBD

Kosaka et al. were the first to emphasise the importance of cortical involvement by Lewy bodies in a post-mortem examination of patients with dementia and Parkinson's disease (Kosaka, 1978). Patients with Lewy bodies were classified in three groups: diffuse (DLBD), transitional and brain stem according to their distribution and number in the central nervous system (Kosaka et al., 1984). Lewy bodies were mainly found in the frontal and temporal lobe, insular cortex and cingulate gyrus. To make a pathological diagnosis of DLBD, these authors require the presence of more than five cortical Lewy bodies per high-powered field in the neocortical areas (Kosaka et al., 1984). In a further review of 37 cases published in the Japanese literature, DLBD was classified into a common form, with presence of Lewy bodies as well as Alzheimer pathology, and a pure form with minimal associated senile changes (Kosaka, 1990). Retrospective analysis of the clinical picture of these patients showed that the common form of DLBD was characterised by progressive cortical dementia appearing early and parkinsonism developing later in 2/3 of the patients while the group with a pure form of DLBD showed predominantly parkinsonian symptoms associated with profound dementia in later stages. Interestingly, 30% of the patients with the common form did not show parkinsonian signs, while conversely, dementia was absent in one out of nine cases with the pure form (Kosaka, 1990). The majority of cases described by the literature as DLBD correspond, pathologically, to the common form (Hansen and Galasko, 1995; Kosaka, 1990).

Post-mortem studies in patients with PD were initially reported to show occasional Lewy bodies in the cerebral cortex of 10% of the brains examined (Jellinger and Grisold, 1982). Greater awareness and use of ubiquitin antibody staining, a more sensitive technique, has produced mounting evidence that cortical Lewy bodies are present in nearly all patients with PD (Hughes et al., 1993; Schmidt et al., 1991; Perry et al., 1991) and suggest that isolated brain stem Lewy body disease may be exceedingly rare. A recent neuropathological study in 100 patients clinically diagnosed as PD, with or without dementia, found brain stem and cortical Lewy bodies in all brains examined (Hughes et al., 1993). Only four of these cases however, fulfilled pathological criteria for DLBD and three of these four patients were demented. Interestingly, in half of the population of PD patients who were noted to have dementia there were no pathological changes of Alzheimer's type. Although these and other findings suggest that cortical Lewy bodies may be responsible for dementia in PD (Hughes et al., 1993; Kosaka, 1990; Gibb et al., 1985), significant numbers of cortical Lewy bodies can be found in non-demented PD patients (Hughes et al., 1993). It has been estimated that 10% of normal controls over the age of 80 have some degree of cortical Lewy body involvement (Gibb and Luthert, 1994). It can be seen, therefore, that while

Parkinson-dementia is a clinical syndrome, the pathological correlate may be AD, PD, or DLBD.

Clinically, patients with pathological evidence of DLBD present with variable degrees of parkinsonism, dementia and psychiatric features (Hansen et al., 1990; Byrne et al., 1989). It has been suggested that the dementia of the Lewy body type may differ from AD in that it shows fluctuating severity (Byrne et al., 1989; McKeith et al., 1992) and greater impairment of frontal function (Hansen et al., 1990). Additionally, DLBD patients may be more prone to develop hallucinations and more sensitive to neuroleptics (McKeith et al., 1992). Where parkinsonism is present, there may be atypical features such as early gait involvement, poor L-dopa response, or supranuclear gaze palsy (Perry et al., 1989; Crystal et al., 1990; Fearnley et al., 1991). The only clinical criteria available at present for diagnosing DLBD apply to patients with dementia (McKeith et al., 1996) and so, do not cover those cases that may have DLBD and normal cognition.

Metabolic studies in vivo

Positron emission tomography (PET) and single photon emission tomography (SPECT) have been extensively used to explore, in vivo, the metabolic and blood-flow changes associated with dementia. PET studies have been performed either with $^{15}O_2$ and $H_2^{15}O$ or with ^{18}F-fluorodeoxyglucose (FDG). SPECT studies of regional cerebral blood-flow have generally utilised [99mTc]-HMPAO.

Alzheimer's disease

PET studies in patients with established AD show global reductions of cerebral metabolism, particularly in parietal and temporal association cortices, with relative sparing of the cerebellum, primary motor and visual cortex, and basal ganglia (Chase et al., 1984; Friedland et al., 1983; Frackowiak et al., 1981; Alavi et al., 1986). The isolated temporo-parietal hypometabolism is characteristically seen in patients with mild cognitive deficits or predominant perceptual problems and constructional apraxia, while those with advanced disease or dysphasia show associated prefrontal hypometabolism (Kennedy et al., 1995; Frackowiak et al., 1981; Foster et al., 1983). In general, the localisation of the cortical hypometabolism correlates well with the magnitude and type of cognitive deficit (Kennedy et al., 1995; Frackowiak et al., 1981; Foster et al., 1983; Friedland et al., 1983). This typical pattern of cortical hypometabolism can also, occasionally, be seen subclinically in at-risk relatives of familial Alzheimer's disease patients (Kennedy et al., 1995).

^{18}F-dopa PET studies in patients with AD and associated extrapyramidal rigidity have shown mean striatal dopaminergic uptake within the normal range, significantly higher than that found in patients with PD, though some individuals have mildly reduced levels of putamen tracer uptake (Tyrrell et al., 1990). In keeping with this finding, a post-mortem analysis in AD patients with parkinsonian symptoms showed normal striatal tyrosine hydroxylase activity (Murray et al., 1995).

SPECT studies with [99mTc]-HMPAO show a pattern of cerebral hypoperfusion that is similar to the pattern of hypometabolism described with PET, and which also correlates with the type and severity of cognitive deficit (Costa et al., 1988; Smith et al., 1988). SPECT, in general however, is a less sensitive technique than PET for detecting altered cerebral function in dementia.

Parkinson's disease and dementia

PET studies of resting glucose metabolism in early hemiparkinsonian patients have generally shown increased metabolism in the contralateral basal ganglia (Miletich et al., 1988), though one study reported decreases (Perlmutter and Raichle, 1985). As the disease generalises the raised lentiform glucose metabolism may normalise (Eidelberg et al., 1990), but covariance analysis shows an abnormal inverse relationship between striatal and frontal function (Eidelberg et al., 1994).

In non-demented PD patients, PET studies have shown either normal cortical blood flow and glucose metabolism (Otsuka et al., 1991; Goto et al., 1993; Karbe et al., 1992), or a widespread reduction to levels between the means observed for controls and demented PD groups (Kuhl et al., 1984b; Okada et al., 1989). Kuhl et al., studied a mixed group of demented and non-demented parkinsonians and found a global reduction in cerebral glucose utilisation that correlated inversely with both the severity of bradykinesia and dementia (Kuhl et al., 1984a). With increasing cognitive impairment, parietal cortex showed a regional metabolic deficit similar to that observed in patients with AD (Kuhl et al., 1984a; Peppard et al., 1988; Karbe et al., 1992; Peppard et al., 1990). One patient with parkinsonism and dementia who showed reduced glucose metabolism in parietal, temporal, prefrontal and premotor areas, subsequently underwent post-mortem examination. He was found to have brain stem, though no cortical, Lewy bodies and no associated AD changes, confirming that this pattern of metabolic deficit is not specific for AD (Schapiro et al., 1990).

SPECT has shown normal cerebral blood flow in non-demented PD patients, while in demented patients there is a reduction of flow with two predominant patterns, either an isolated frontal or a fronto-temporo-parietal hypoperfusion. Interestingly, a history of nocturnal delirium was often found in those patients with fronto-parietal but not isolated frontal involvement (Sawada et al., 1992).

Recent clinicopathological studies have suggested that it may be possible to clinically distinguish the cognitive deficits and parkinsonism seen in DLBD from those seen in AD by applying a set of criteria: fluctuating but persistent cognitive impairment, hallucinations and extrapyramidal features (McKeith et al., 1996). We have studied with ^{18}FDG and PET, three cases with cognitive disorder and parkinsonism that fulfilled these criteria.

Case 1

A 71-year-old right-handed man first noticed progressive memory loss five years earlier followed by impaired verbal fluency and confabulation. More recently, he developed tremor of the right arm. On examination at the time of his PET scan he showed resting and postural tremor of the right limbs associated with increased rigidity, bradykinesia and decreased arm-swing of the right hand. His MiniMental scale (MMSE) score was 27/30. He had formal psychometric testing performed twice over three years; the second testing, a few months before PET, revealed progressive impairment of frontal function but improvement in his performance IQ. His brain CT showed involutional changes without focal abnormalities. At the time of the PET study he was not receiving any medication.

Case 2

A 72-year-old right-handed woman presented with a two year history of a shuffling gait. Diagnosed as having Parkinson's disease, she was started on an L-dopa/carbidopa preparation with immediate improvement of her symptoms. After a few months she developed end-dose akinesia and peak dose dyskinesias in response to L-dopa. Two years later she noticed a deterioration of her memory. At the time of the PET study she was experiencing depression, geographical and temporal disorientation, confusion in the mornings and visual hallucinations (mainly in the evenings). On examination while off medication, she had hypophonia, a bilateral resting tremor, mild rigidity of the right side and decreased arm-swing bilaterally. She had a shuffling gait and impaired postural reflexes. Her MMSE was 22/30 and on bedside testing there was a deficit of recent recall, attention and praxis. Her brain CT showed basal ganglia calcification. At the time of the PET study she was receiving 400 mg of controlled release L-dopa/carbidopa, fluoxetine 20 mg and diazepam 5 mg daily.

Case 3

A 73-year-old right-handed man had a 13 year history of action tremor and a two year history of occasional geographical disorientation, gradual loss of memory, and an episode of visual hallucinations. At the time of PET he was complaining of fluctuating memory disturbance and depression. On examination he had hypophonia, mild bilateral resting and action tremor, absent arm-swing and bradykinesia most noticeable in the right hand. He had a stooped posture, a shuffling gait and impaired postural reflexes. The MMSE was 26/30 and he exhibited attentional deficit and apraxia. His brain CT showed mild involutional changes without focal abnormalities. At the time of the PET study he was receiving selegiline 10 mg/day.

All three patients received an intravenous bolus of 5 mCi [18]FDG and were scanned for 60 minutes. Results were individually compared with those obtained for a group of 15 age-matched normal controls using statistical parametric mapping (SPM) as described elsewhere to demonstrate focal reductions in glucose metabolism (Frackowiak and Friston, 1994; Friston, 1995). Differences in global glucose metabolism (CMRglc) between subjects were normalised using an analysis of covariance (ANCOVA) with global metabolism as the confounding variable. Focal glucose metabolism for each individual was then compared on a pixel-by-pixel basis with that of a group of normals applying Z scores to each pixel. Due to the preliminary character of this study the results are shown for those voxels in which the reduction of glucose metabolism was greater than a Z score of three.

Patient 1

Decreases in rCMRglc were most pronounced in bilateral mesial and lateral prefrontal cortex (areas 8 and 9), inferior temporal cortex, hippocampal gyrus, and right parietal area 40.

Patient 2

Decreases in rCMRglc were found in bilateral dorsal prefrontal area 9, anterior supplementary motor area and anterior cingulate, inferior temporal cortex, hippocampal gyrus, and bilateral parietal area 40.

Patient 3

Decreases in rCMRglc were seen in left mesial and lateral prefrontal cortex (areas 8 and 9), anterior supplementary motor area, Broca's area 44, posterior cingulate/retrosplenial cortex (23), and bilateral inferior temporal cortex.

These preliminary results in our three patients who fulfilled clinical diagnostic criteria for DLBD, showed a reduction of glucose metabolism in fronto-temporo-parietal areas, similar to that previously described in patients with AD and PD-dementia. It appears that when cognitive failure is established the overall pattern of metabolic dysfunction is similar. Although these three patients only had mild cognitive failure they all showed a frontal metabolic deficit, while in AD patients with mild dementia the metabolic deficit is often restricted to temporo-parietal areas (Kennedy et al., 1995; Frackowiak et al., 1981; Foster et al., 1983; Friedland et al., 1983). The only reported patient with pathological confirmed PD and mild dementia also showed glucose hypometabolism in fronto-temporo-parietal areas (Schapiro et al., 1990).

Two other extrapyramidal disorders also have associated cognitive deficits: progressive supranuclear palsy (PSP) and corticobasal degeneration (CBD). The impairment of cognition is different from that observed in AD, or PD-dementia, with a progressive slowness of intellect in PSP and predominant apraxia and agnosia in CBD.

Progressive supranuclear palsy

This rapidly progressive akinetic-rigid syndrome is associated with a supranuclear gaze palsy, axial dystonia, dysarthria, dysphagia and cognitive dysfunction of "subcortical" type (Litvan, 1994; Duvoisin, 1994; Colosimo et al., 1995). Patients generally have a poor recall, impaired fluency and planning, but a more generalised cognitive deficit including agnosia and aphasia is most unusual (Duvoisin, 1994; Tolosa et al., 1994). The pathology consists of neuronal loss, neurofibrillary tangles and gliosis in the brain stem nuclei, striatum and globus pallidus, with involvement of frontal cortex in 40% of the cases (De Bruin and Lees, 1994). PET and SPECT studies in PSP patients, some of them with post-mortem diagnostic confirmation (Foster et al., 1992), have shown global reduction of rCMRglc and hypoperfusion, most pronounced in frontal areas and the basal ganglia — [See review Brooks (1994)], (Karbe et al., 1992; Johnson et al., 1992). Blin et al. reported a correlation between cognitive performance and degree of fronto-occipital hypometabolism, as well as an inverse correlation between motor scores and basal ganglia metabolism (Blin et al., 1990a). This pattern of resting metabolism is quite distinct from the reduced fronto-temporo-parietal and normal striatal metabolism observed in patients with AD and PD-dementia. It can also be found in patients with striatonigral degeneration, although dementia is rare in this condition (Goffinet et al., 1989).

Striatal ^{18}F-dopa uptake is symmetrically reduced in PSP (Golbe et al., 1989). Caudate and putamen show an equivalent loss of dopaminergic terminal function (Burn et al., 1994), in contrast to PD, in which caudate function is relatively preserved.

Corticobasal degeneration

This is an asymmetric akinetic-rigid syndrome poorly L-dopa responsive, characterised by limb apraxia and myoclonus, a supranuclear gaze palsy, and bulbar dysfunction. Occasionally, patients may exhibit alien limb behaviour. Memory disturbance in this condition is generally mild and only appears at a late stage (Rinne et al., 1994; Rebeiz et al., 1968). The pathology reveals neuronal achromasia and atrophy in the fronto-temporo-parietal cortex and loss of neurons and gliosis in the pars compacta of the substantia nigra and other subcortical structures (Rebeiz et al., 1968; Watts et al., 1994). PET studies have shown strikingly asymmetric resting hypometabolism of the pathologically affected cortical areas and the thalamus (Sawle et al., 1991; Blin et al., 1990b, 1992). Resting rCMRglc asymmetries are significantly greater than those observed in patients with asymmetric PD (Eidelberg et al., 1991), and together with thalamic involvement, this pattern differentiates CBD from AD and PD-dementia. Striatal ^{18}F-dopa uptake is asymmetrically reduced, and caudate and putamen involvement are equivalent (Sawle et al., 1991).

In summary, parkinsonism-dementia may be a clinical expression of either isolated brain stem or diffuse Lewy body disease or a combination of Lewy body and AD pathologies. PET studies in patients with parkinsonism-dementia have shown a pattern of fronto-temporo-parietal hypometabolism indistinguishable from that observed in patients with AD, even when only isolated brain stem Lewy body disease was subsequently found at pathology. ^{18}FDG PET in three of our PD patients with mild cognitive dysfunction that fulfilled diagnostic criteria for DLBD also showed an AD metabolic pattern. Further PET studies are required to determine whether early frontal hypometabolism may help to distinguish DLBD from AD. Two other neuro-degenerative conditions may present as parkinsonism associated with a cognitive deficit: PSP where cognitive dysfunction is mainly frontal and CBD where it is mild and appears at a late stage. Metabolic studies in PSP have shown global reduction of rCMRglc most pronounced in frontal areas and the basal ganglia while in CBD there is a cortical and thalamic hypometabolism contralateral to the affected limbs. These results suggest that while PET studies of resting metabolism may help to distinguish PD-dementia, PSP and CBD, they are not helpful in distinguishing PD-dementia from AD, irrespective of the pathological substrate of the former.

References

Alavi A, Dann R, Chawluk J, Alavi J, Kushner M, Reivich M (1986) Positron emission tomography imaging of regional cerebral glucose metabolism. Semin Nucl Med 16: 2–34

Blin J, Baron JC, Dubois B, Pillon B, Cambon H, Cambier J, Agid Y (1990a) Positron emission tomography study in progressive supranuclear palsy. Brain hypometabolic pattern and clinicometabolic correlations. Arch Neurol 47: 747–752

Blin J, Vidailhet M, Bonnet AM, Dubois B, Pillon B, Syrota A, Agid Y (1990b) PET study in cortico-basal degeneration. Mov Disord 5 [Suppl 1]: 19

Blin J, Vidailhet M-J, Pillon B, Dubois B, Feve J-R, Agid Y (1992) Corticobasal degeneration: decreased and asymmetrical glucose consumption as studied with PET. Mov Disord 7: 348–354

Brooks DJ (1994) PET studies in progressive supranuclear palsy. J Neural Transm [Suppl 42]: 119–134

Brown RG, Marsden CD (1984) How common is dementia in Parkinson's disease? Lancet ii: 1262–1265

Burn DJ, Sawle GV, Brooks DJ (1994) Differential diagnosis of Parkinson's disease, multiple system atrophy, and Steele-Richardson-Olszewski syndrome: discriminant analysis of striatal ^{18}F-dopa PET data. J Neurol Neurosurg Psychiatry 57: 278–284

Byrne J, Lennox G, Lowe J, Godwin-Austen RB (1989) Diffuse Lewy body disease: clinical features in 15 cases. J Neurol Neurosurg Psychiatry 52: 709–717

Chase TN, Foster NL, Fedio P, Brooks R, Mansi L, Di Chiro G (1984) Regional cortical dysfunction in Alzheimer's disease as determined by positron emission tomography. Ann Neurol 15 [Suppl]: S170–S174

Colosimo C, Albanese A, Hughes AJ, De Bruin VMS, Lees AJ (1995) Some specific clinical features differentiate multiple system atrophy (striatonigral variety) from Parkinson's disease. Arch Neurol 52: 294–298

Costa DC, Ell PJ, Philpot M, Levy R (1988) CBF tomograms with ^{99m}Tc-HM-PAO in patients with dementia (Alzheimer's type and HIV) and Parkinson's disease-initial results. J Cereb Blood Flow Metab 8 [Suppl 1]: S109–S115

Crystal HA, Dickson DW, Lizardi JE, Davies P, Wolfson LI (1990) Antemortem diagnosis of diffuse Lewy body disease. Neurology 40: 1523–1528

De Bruin VMS, Lees AJ (1994) Subcortical neurofibrillary degeneration presenting as Steele-Richardson-Olszewski and other related syndromes: a review of 90 pathologically verified cases. Mov Disord 9: 381–389

Ditter SM, Mirra SS (1987) Neuropathologic and clinical features of Parkinson's disease in Alzheimer's disease patients. Neurology 37: 754–760

Duvoisin RC (1994) Differential diagnosis of PSP. J Neural Transm [Suppl 42]: 51–67

Eidelberg D, Moeller JR, Dhawan V, Sidtis JJ, Ginos JZ, Strother, SC, Cedarbaum J, Greene P, Fahn S, Rottenberg DA (1990) The metabolic anatomy of Parkinson's disease: complementary [18F]Fluorodeoxyglucose and [18F]Fluorodopa positron emission tomographic studies. Mov Disord 5: 203–213

Eidelberg D, Dhawan V, Moeller JR, Sidtis JJ, Ginos JZ, Strother SC, Cederbaum J, Greene P, Fahn S, Powers JM, Rottenberg DA (1991) The metabolic landscape of cortico-basal ganglionic degeneration: regional asymmetries studied with positron emission tomography. J Neurol Neurosurg Psychiatry 54: 856–862

Eidelberg D, Moeller JR, Dhawan V, Spetsieris P, Takikawa S, Ishikawa T, Chaly T, Robeson W, Margouleff D, Przedborski S, Fahn S (1994) The metabolic topography of parkinsonism. J Cereb Blood Flow Metab 14: 783–801

Fearnley JM, Revesz T, Brooks DJ, Frackowiack RSJ, Lees AJ (1991) Diffuse Lewy body disease presenting with a supranuclear gaze palsy. J Neurol Neurosurg Psychiatry 54: 159–161

Foster NL, Chase TN, Fedio P, Patronas NJ, Brooks RA, Di Chiro G (1983) Alzheimer's disease: focal cortical changes shown by positron emission tomography. Neurology 33: 961–965

Foster NL, Gilman S, Berent S, Sima AAF, D'Amato C, Koeppe RA, Hicks SP (1992) Progressive subcortical gliosis and progressive supranuclear palsy can have similar clinical and PET abnormalities. J Neurol Neurosurg Psychiatry 55: 707–713

Frackowiak RSJ, Friston KJ (1994) Functional neuroanatomy of the human brain: positron emission tomography — a new neuroanatomical technique. J Anat 184: 211–225

Frackowiak RSJ, Pozilli C, Legg, NJ, Du Boulay GH, Marshall J, Lenzi GL, Jones T (1981) Regional cerebral oxygen supply and utilization in dementia. Brain 104: 753–778

Friedland RP, Budinger TF, Ganz E, Yano Y, Mathis CA, Koss B, Ober BA, Huesman RH, Derenzo SE (1983) Regional cerebral metabolic alterations in dementia of the Alzheimer type: positron emission tomography with [18F]Fluorodeoxyglucose. J Comput Assist Tomogr 7: 590–598

Friston KJ (1995) Statistical parametric mapping: ontology and current issues. J Cereb Blood Flow Metab 15: 361–370

Gibb WRG, Luthert PJ (1994) Dementia in Parkinson's disease and Lewy body disease. In: Burns A, Levy R (eds) Dementia. Chapman & Hall, London, pp 719–738

Gibb WRG, Esiri MM, Lees AJ (1985) Clinical and pathological features of diffuse cortical Lewy body disease (Lewy body dementia). Brain 110: 1131–1153

Goffinet AM, De Volder AG, Gillain C, Rectem D, Bol A, Michel C, Cogneau M, Labar D, Laterre C (1989) Positron tomography demonstrates frontal lobe hypometabolism in progressive supranuclear palsy. Ann Neurol 25: 131–139

Golbe LI, Miller DC, Duvoisin RC (1989) Paraneoplastic degeneration of the substantia nigra with dystonia and parkinsonism. Mov Disord 4: 147–152

Goto I, Taniwaki T, Hosokawa S, Otsuka M, Ichiya Y, Ichimiya A (1993) Positron emission tomographic (PET) studies in dementia. J Neurol Sci 114: 1–6

Hansen L, Salmon D, Galasko D, Mesliah E, Katzman R, DeTeresa R, Thal L, Pay MM, Hofstetter R, Klauber M, Rice V, Butters N, Alford M (1990) The Lewy body variant of Alzheimer's disease: a clinical and pathologic entity. Neurology 40: 1–8

Hansen LA, Galasko D (1995) Lewy body disease. Curr Opin Neurol Neurosurg 5: 889–894

Hughes AJ, Daniel SE, Blankson S, Lees AJ (1993) A clinicopathologic study of 100 cases of Parkinson's disease. Arch Neurol 50: 140–148

Jellinger K, Grisold W (1982) Cerebral atrophy in Parkinson syndrome. Exp Brain Res [Suppl] 5: 26–35

Johnson KA, Sperling RA, Holman BL, Nagel JS, Growdon JH (1992) Cerebral perfusion in progressive supranuclear palsy. J Nucl Med 33: 704–709

Karbe H, Holthoff V, Huber M, Herholz Z, Wienhard K, Wagner R, Heiss WD (1992) Positron emission tomography in degenerative disorders of the dopaminergic system. J Neural Transm [PD Sect] 4: 121–130

Kennedy AM, Frackowiak RSJ, Newman SK, Bloomfield PM, Seaward J, Roques P, Lewington G, Cunningham VJ, Rossor MN (1995) Deficits in cerebral glucose metabolism demonstrated by positron emission tomography in individuals at risk of familial Alzheimer's disease. Neurosci Lett 186: 17–20

Kosaka K (1978) Lewy bodies in cerebral cortex. Report of three cases. Acta Neuropathol 42: 127–134

Kosaka K (1990) Diffuse Lewy body disease in Japan. J Neurol 237: 197–204

Kosaka K, Yoshimura M, Ikeda K, Budka H (1984) Diffuse type of Lewy body disease: progressive dementia with abundant cortical Lewy bodies and senile changes of various degree — a new disease? Clin Neuropathol 3: 185–192

Kuhl DE, Metter EJ, Riege WH (1984a) Patterns of local cerebral glucose utilization determined in Parkinson's disease by the [18F]Fluorodeoxyglucose method. Ann Neurol 15: 419–424

Kuhl DE, Metter EJ, Riege WH, Markham CH (1984b) Patterns of cerebral glucose utilization in Parkinson's disease and Huntington's disease. Ann Neurol 15 [Suppl]: S119–S125

Lees AJ (1985) Parkinson's disease and dementia. Lancet i: 43–44

Litvan I (1994) Cognitive disturbances in progressive supranuclear palsy. J Neural Transm [Suppl 42]: 69–78

Mayeux R, Stern Y, Spanton S (1985) Heterogeneity in dementia of the Alzheimer's type: evidence of subgroups. Neurology 35: 453–461

McKeith I, Fairbairn A, Perry R, Thompson P, Perry E (1992) Neuroleptic sensitivity in patients with senile dementia of Lewy body type. BMJ 305: 673–678

Mc Keith IG, Galesko D, Kosaka K, Perry EK, Dickson DW, Hansen LA, Salmon DP, Lowe J, Mirra SS, Byrne EJ, Lennox G, Quinn NP, Edwardson JA, Ince PG, Bergeron C, Burns A, Miller BL, Lovestone S, Collerton D, Jansen ENH, Ballard C, de Vos RAI, Wilcock GK, Jellinger KA, Perry RH for the Consortium on Dementia with Lewy Bodies (1996) Consensus guidelines for the clinical and pathologic diagnosis of dementia with Lewy bodies (DLB): report of the consortium on DLB international workshop. Neurology 47: 1113–1124

Miletich RS, Chase T, Gillespie M, Di Chiro G, Stein S (1988) Contralateral basal ganglia glucose metabolism is abnormal in hemi-parkinsonian patients: an FDG-PET study. Neurology 38 [Suppl 1]: 260

Molsa PK, Marttila RJ, Rinne UK (1984) Extrapyramidal signs in Parkinson's disease. Neurology 34: 1114–1116

Murray AM, Weihmueller FB, Marshall JF, Hurtig HI, Gottleib GL, Joyce JN (1995) Damage to dopamine systems differs between Parkinson's disease and Alzheimer's disease with parkinsonism. Ann Neurol 37: 300–312

Okada J, Peppard R, Calne DB (1989) Comparison study of positron emission tomography, X-ray CT and MRI in parkinsonism with dementia. Nippon Igaku Hoshasen Gakkai Zasshi 49: 643–656

Otsuka M, Ichiya Y, Hosokawa S, Kuwabara Y, Tahara T, Fukumura T, Kato M, Masuda K, Goto I (1991) Striatal blood flow, glucose metabolism and ^{18}F-dopa uptake: difference in Parkinson's disease and atypical parkinsonism. J Neurol Neurosurg Psychiatry 54: 898–904

Peppard RF, Martin WRW, Guttman M, McGeer PL, Walsh EM, Carr GD, Phillips AG, Grochowski E, Okada J, Tsui JKC, Mak E, Ruth E, Adam MJ, Calne DB (1988) The relationship of cerebral glucose metabolism to cognitive deficits in Parkinson's disease. Neurology 38 [Suppl 1]: 364

Peppard RF, Martin WR, Clark CM, Carr GD, McGeer PL, Calne DB (1990) Cortical glucose metabolism in Parkinson's and Alzheimer's disease. J Neurosci Res 27: 561–568

Perlmutter JS, Raichle ME (1985) Regional blood flow in hemiparkinsonism. Neurology 35: 1127–1134

Perry EK, McKeith I, Thompson P, Marshall E, Kerwin J, Jabeen S, Edwardson JA, Ince P, Blessed G, Irving D, Perry RH (1991) Topography, extent, and clinical relevance of neurochemical deficits in dementia of Lewy body type, Parkinson's disease, and Alzheimer's disease. Ann NY Acad Sci 640: 197–202

Perry RH, Irving D, Blessed G, Perry EK, Fairbairn AF (1989) Senile dementia of Lewy body type and spectrum of Lewy body disease. Lancet i: 1088

Rebeiz JJ, Kolodny EH, Richardson EP (1968) Corticodentatonigral degeneration with neuronal achromasia. Arch Neurol 18: 20–33

Rinne JO, Lee MS, Thompson PD, Marsden CD (1994) Corticobasal degeneration. A clinical study of 36 cases. Brain 117: 1183–1196

Sawada H, Udaka F, Kameyama M, Seriu N, Nishinaka K, Shindou K, Kodama M, Nishitani N, Okumiya K (1992) SPECT findings in Parkinson's disease associated with dementia. J Neurol Neurosurg Psychiatry 55: 960–963

Sawle GV, Brooks DJ, Marsden CD, Frackowiak RSJ (1991) Corticobasal degeneration. A unique pattern of regional cortical oxygen hypometabolism and striatal fluorodopa uptake demonstrated by positron emission tomography. Brain 114: 541–556

Schapiro MB, Grady C, Ball MJ, DeCarli C, Rapoport SI (1990) Reductions in parietal/temporal cerebral glucose metabolism are not specific for Alzheimer's disease. Neurology 40 [Suppl 1]: 152

Schmidt ML, Murray J, Lee VM-Y, Hill WD, Wertkin A, Trojanowski JQ (1991) Epitope map of neurofilament protein domains in cortical and peripheral nervous system Lewy bodies. Am J Pathol 139: 53–65

Smith FW, Besson JAO, Gemmell HG, Sharp PF (1988) The use of Technetium-99m-HM-PAO in the assessment of patients with dementia and other neuropsychiatric conditions. J Cereb Blood Flow Metab 8 [Suppl 1]: S116–S122

Tison F, Dartigues JF, Auriacombe S, Letenneur L, Boller F, Alperovitch A (1995) Dementia in Parkinson's disease: a population-based study in ambulatory and institutionalized individuals. Neurology 45: 705–708

Tolosa E, Valldeoriola F, Marti MJ (1994) Clinical diagnosis and diagnostic criteria of progressive supranuclear palsy (Steele-Richardson-Olszewski syndrome). J Neural Transm 42 [Suppl]: 15–31

Tyrrell PJ, Sawle GV, Ibanez V, Bloomfield PM, Leenders KL, Frackowiak RS, Rossor MN (1990) Clinical and positron emission tomographic studies in the "extrapyramidal syndrome" of dementia of the Alzheimer type. Arch Neurol 47: 1318–1323

Watts RL, Mirra SS, Richardson EP (1994) Corticobasal ganglionic degeneration. In: Marsden CD, Fahn S (eds) Movement disorders 3. Butterworth-Heinemann, London, pp 282–299

Authors' address: Dr. N. Turjanski, MRC Cyclotron Unit, Hammersmith Hospital, DuCane Road, London W12 OHS, United Kingdom

Neurofibrillary tangles and neuropil threads as a cause of dementia in Parkinson's disease*

H. Braak[1], **E. Braak**[1], **D. Yilmazer**[1], **R. A. I. de Vos**[2], **E. N. H. Jansen**[2], and **J. Bohl**[3]

[1] Zentrum der Morphologie, Universität Frankfurt/Main, Federal Republic of Germany
[2] Streeklaboratoria voor pathologie, Enschede, The Netherlands
[3] Abteilung für Neuropathologie, Universität Mainz, Federal Republic of Germany

Summary. Alzheimer's disease (AD) and Parkinson's disease (PD) are the most common age-related degenerative disorders of the human brain. Both diseases involve multiple neuronal systems and are the consequences of cytoskeletal abnormalities. In AD susceptible neurons produce neurofibrillary changes, while in Parkinson's disease, they develop Lewy bodies. In AD six developmental stages can be distinguished on account of the predictable manner in which the neurofibrillary changes spread across the cerebral cortex. During the course of PD numerous limbic determined parts of the brain undergo specific lesions regulating endocrine and autonomic functions. In general, the extranigral destructions are in themselves not sufficient to produce overt intellectual deterioration. Fully developed Parkinson's disease with concurring incipient Alzheimer's disease is likely to cause impaired cognition.

Alzheimer's disease (AD) and Parkinson's disease (PD) are the most common age-related degenerative disorders of the human brain. Both diseases are the consequences of cytoskeletal abnormalities developing in only a few neuronal types. In AD, susceptible neurons produce neurofibrillary tangles (NFTs) and neuropil threads (NTs) (Braak et al., 1994a), while in PD, they develop Lewy bodies (LBs) and Lewy neurites (LNs). Neurons displaying the specific changes eventually die for as yet unknown reasons (Braak et al., 1994b, 1995, 1996; Fearnley and Lees, 1994; Gibb and Lees, 1991; Gibb et al., 1991; Jellinger, 1990, 1991, 1994; Lowe, 1994).

Both AD and PD, involve multiple neuronal systems. The specific lesional pattern of these illnesses accrues slowly over time and remains remarkably consistent across cases with only minor interindividual variation. The pathologic process underlying AD, preferentially destroys "afferent" cortical structures such as the entorhinal region and the neocortical association areas.

---

*The results are described in detail elsewhere (Braak et al., 1996)

In PD, in contrast, "efferent" subcortical structures such as the central nucleus of the amygdala and the substantia nigra are the main targets of destruction.

In AD, six developmental stages can be distinguished on account of the predictable manner in which the neurofibrillary changes spread across the various areas of the cerebral cortex (Braak and Braak, 1991). The pathologic process initially destroys a few projection cells of the transentorhinal region (representing the clinically silent transentorhinal stages I and II), then proceeds into other cortical and subcortical components of the limbic system (limbic stages III and IV), and eventually extends into the neocortical association areas (neocortical stages V and VI). Stages III and IV are considered to represent incipient AD, while stages V and VI correspond to fully developed AD (Bancher et al., 1993; Braak et al., 1993; Jellinger et al., 1991).

In the normal human brain, somato-sensory, visual, and auditory data proceed through the respective core and belt regions of the parietal, occipital, and temporal neocortex and are then transferred to a large number of association areas. From here, the data are conveyed by long cortico-cortical projections to the prefrontal association cortex. Part of the stream of data from the

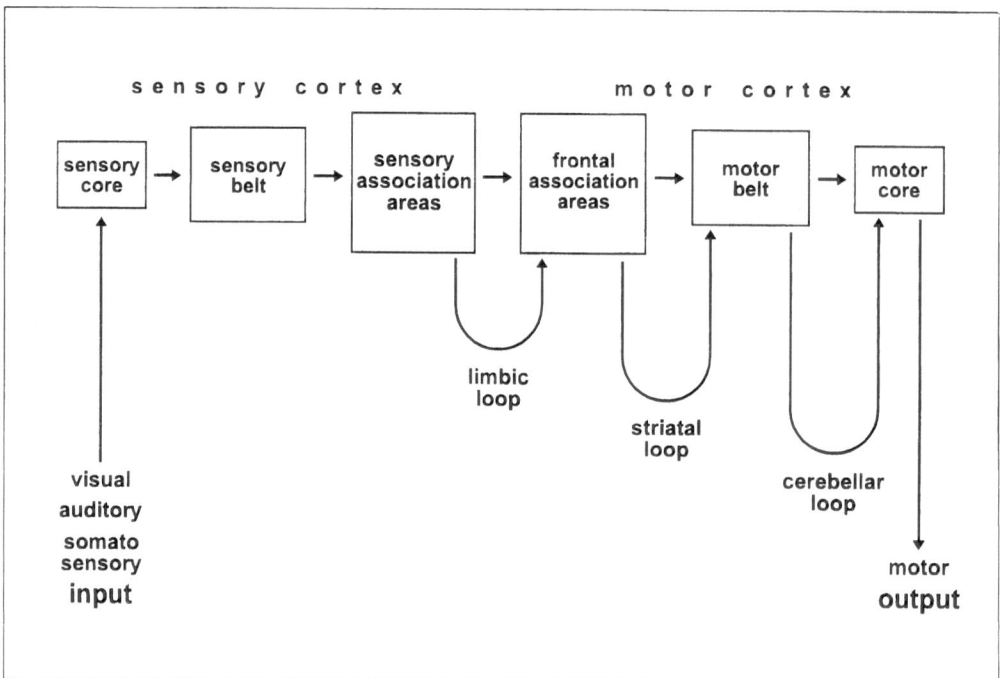

Fig. 1. The parietal, occipital, and temporal neocortex is each comprised of a core field, a belt region, and extensive association areas. Somato-sensory, visual, and auditory information proceeds through core and belt regions to a variety of association areas and then the data is transported via long cortico-cortical projections to the prefrontal association fields. Tracts generated from this highest organisational level of the brain guide the data back via the frontal belt areas to the frontal core field, the primary motor area. The striatal and cerebellar loops provide the major routes for this transport from the prefrontal cortex to the primary motor field

sensory to the motor association areas branches off and converges upon the allocortex and amygdala (Felleman and van Essen, 1991). The sensory neo-cortical data are processed within the relay stations of the limbic loop and are then conveyed to the prefrontal cortex. Tracts generated from the prefrontal areas then guide the data via the frontal belt (premotor areas) to the frontal core (primary motor area). The striatal and the cerebellar loops provide the major routes for this transport from the prefrontal cortex to the primary motor field (Figs. 1 and 2, upper half) (Alexander et al., 1990; Alheid et al., 1990).

Both AD and PD, involve many relay stations in the limbic system. In AD, the major portals of entry for neocortical information, i.e. the transentorhinal region and the lateral nucleus of the amygdala are involved quite early in the course of the disease. At stage III, the entorhinal cortex exhibits severe destruction of both the outer cellular layer, which is responsible for the transfer of neocortical data to the hippocampus, and one of the inner cellular layers responsible for the feed-back projection from the subiculum of the hippocampal formation to the neocortex. Stage III destruction of the amygdala is also associated with a reduction of data directly entering the CA1 sector of the hippocampus. At this stage, additional lesions are seen in the thalamic midline nuclei, the anterodorsal thalamic nucleus, and the retrosplenial region (Braak and Braak, 1993). Taken together, the lesions at stage III partially or totally disrupt important limbic circuits at multiple sites. A few patients with stage III destruction and many at stage IV exhibit memory dysfunction and initial signs of cognitive impairment (Bancher et al., 1993; Braak et al., 1993). Obviously, the specific bilateral lesions of a few key components of the limbic system contribute to the development of the char-acteristic symptoms of AD (Fig. 2, lower half).

In PD, one of the inner layers of the entorhinal region (layer Pri-α) is exclusively affected. Large numbers of LBs occur in this layer which is respon-sible for the feed-back projection of hippocampal data to the neocortex (Goldman-Rakic et al., 1984; Kosel et al., 1982). The hippocampus shows a wealth of LNs in the CA2 sector (Braak et al., 1994b; Dickson et al., 1991). The midline nuclei of the thalamus are particularly severely involved. In addition, there is a remarkable change in the anterior pro-neocortex, includ-ing the anterior fields of the cingulate gyrus, anterobasal insula and temporal areas adjoining the allocortical core. Abundant LBs are seen in these areas, in particular within the multiform layer. The brunt of the PD-related changes is born by the amygdala (Braak et al., 1994b). The ventromedial divisions of the basal and the accessory basal nucleus, as well as the accessory cortical nucleus and, in particular, the central nucleus of the amygdala exhibit dense accumu-lations of small LBs and LNs. The amygdala not only contributes to the efferent leg of the limbic loop but also sends important projections to many nuclei exerting influence upon the regulation of endocrine and autonomic functions. Furthermore, it controls all non-thalamic nuclei which project in a non-specific manner to the cerebral cortex (i.e. the cholinergic nuclei of the basal forebrain, the histaminergic tuberomamillary nucleus, the dopaminergic nuclei of the ventral tegmentum, the serotonergic nuclei of the anterior raphe

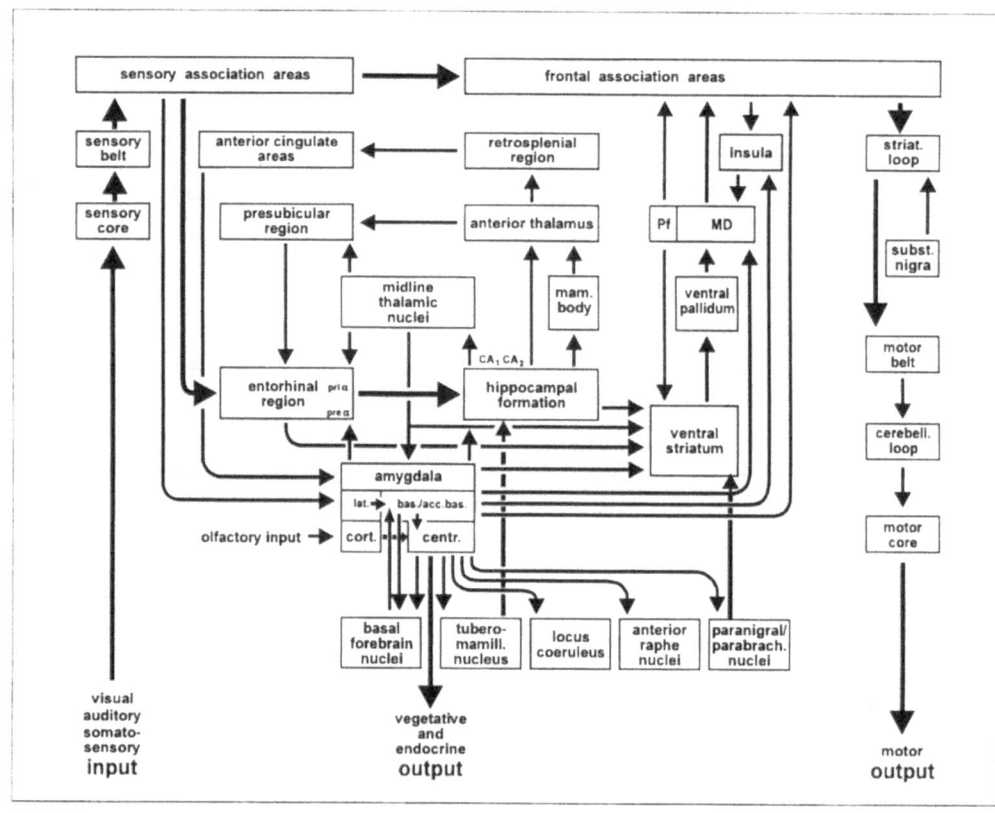

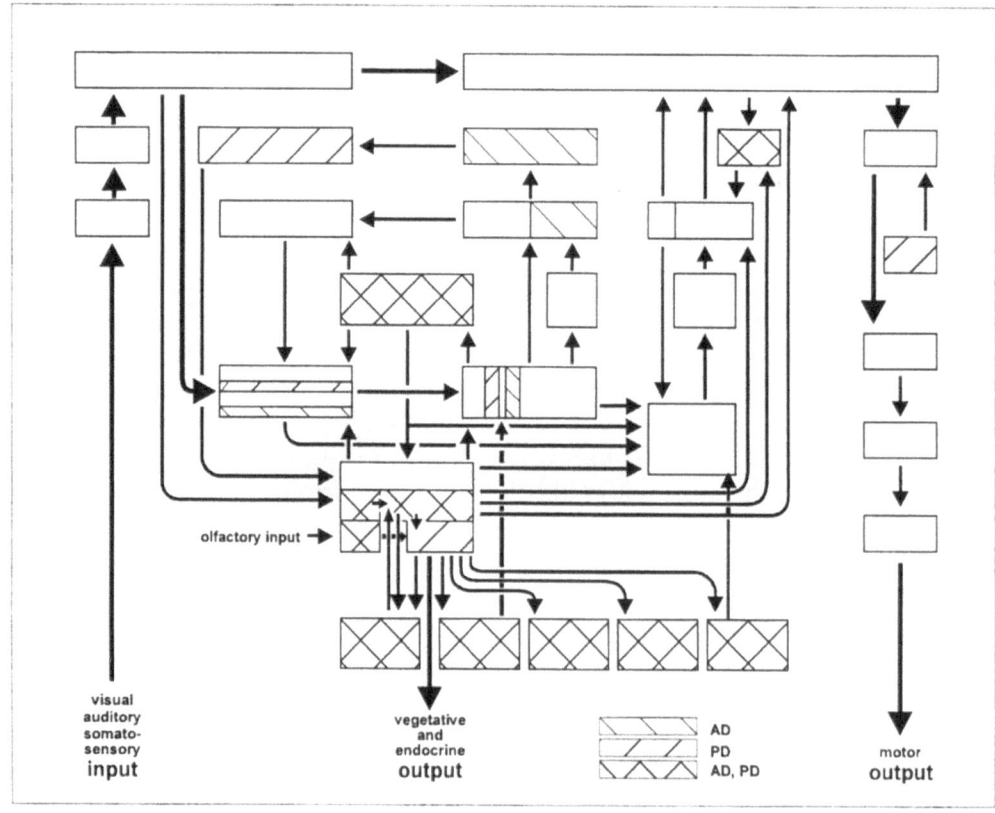

complex, and the noradrenergic locus coeruleus). In PD, all these amygdala-dependent nuclei show marked changes. Hence, apart from the conspicuous pathology of the substantia nigra, there is also a substantial degree of extranigral lesions which gradually result in the destruction of important relay stations in the limbic system. These lesions are in themselves not sufficient to produce overt intellectual impairment but may pave the way for the appearance of cognitive decline (Fig. 2, lower half).

Both disorders, PD and AD, are closely related and they frequently co-occur in the same individual. In such cases of overlap, both disorders maintain their characteristic lesional patterns. Many areas and nuclei show destructions of only one of the two diseases, while others display a combination of PD- and AD-specific lesions.

Neuropathologic evaluation of PD cases with impaired cognition frequently reveals fully developed PD lesions in the presence of AD destruction of stage III or higher while mentally unimpaired cases remain devoid of AD changes or exhibit changes corresponding to no more than stage I or II (Braak and Braak, 1990). In this context, it should be emphasized that the initial stages of AD pathology easily escape recognition. Both in PD and AD, the entorhinal cortex is usually not evaluated in routine neuropathologic examinations. The conventional hippocampus block cut out at the level of the lateral geniculate body cannot be used for evaluation of the pathologic lesions in the entorhinal cortex.

In summary, it appears tempting to suggest that in full blown PD, relatively slight additional lesions in the form of AD stage III destruction produces a potentiating effect that eventually leads to mental deterioration (Bancher et al., 1993; Braak and Braak, 1990; Jellinger and Bancher, 1995; Jellinger et al., 1991). This does not rule out that individual PD cases may present with other causes for dementia such as the co-occurrence of

---

Fig. 2. *Upper half* Part of the stream of data from the sensory association areas to the prefrontal cortex branches off and eventually converges upon the entorhinal region and the amygdala (afferent leg of the limbic loop). The transentorhinal region and the lateral nucleus of the amygdala serve as major gates of entrance for this highly processed neocortical information. Via connections with the hippocampal formation, the neocortical data is then distributed to a variety of related limbic structures. Projections from the hippocampal formation, the entorhinal region, and the amygdala contribute to the efferent leg of the limbic loop which is directed to and exerts important influence upon the prefrontal cortex. The amygdala integrates exteroceptive sensory data with interoceptive stimuli from autonomic centres. A large number of amygdala efferents terminate in nuclei regulating endocrine and autonomic functions. In addition, the amygdala generates efferent connections to all non-thalamic nuclei which in a non-specific manner project upon the cerebral cortex. *Lower half* Drawing of the upper figure showing the various areas involved in Alzheimer's disease (AD) and Parkinson's disease (PD) and in a combination of both. *cerebell. loop* cerebellar loop; *lat., bas., acc.bas., cort., centr.* lateral, basal, accessory-basal, cortical, central nuclei of the amygdala; *mam. body* mamillary body; *MD* mediodorsal nucleus of thalamus; *parabrach.* parabrachial; *PF* parafascicular nucleus; *striat. loop* striatal loop; *subst. nigra* substantia nigra; *tuberomamill. nucleus* tuberomamillary nucleus

argyrophilic grains (Braak and Braak, 1989), of multiple and extended infarctions or other specific lesions. The presently available data, however, support the view that the co-occurrence of AD-related brain destruction is the most common cause of intellectual decline in PD.

Acknowledgements

This study was supported by the Deutsche Forschungsgemeinschaft and the Bundeministerium für Forschung und Technologie. The skilful assistance of Ms. I.Szasz (drawings) is gratefully acknowledged.

References

Alexander GE, Crutcher MD, DeLong MR (1990) Basal ganglia-thalamocortical circuits: parallel substrates for motor, oculomotor, "prefrontal" and "limbic" functions. Prog Brain Res 85: 119–146

Alheid GF, Heimer L, Switzer RC (1990) Basal ganglia. In: Paxinos G (ed) The human nervous system. Academic Press, New York, pp 483–582

Bancher C, Braak H, Fischer P, Jellinger KA (1993) Neuropathological staging of Alzheimer lesions and intellectual status in Alzheimer's and Parkinson's disease. Neurosci Lett 162: 179–182

Braak H, Braak E (1989) Cortical and subcortical argyrophilic grains characterize a disease associated with adult onset dementia. Neuropathol Appl Neurobiol 15: 13–26

Braak H, Braak E (1990) Cognitive impairment in Parkinson's disease: amyloid plaques, neurofibrillary tangles and neuropil threads in the cerebral cortex. J Neural Transm [P-D Sect] 2: 45–57

Braak H, Braak E (1991) Neuropathological stageing of Alzheimer-related changes. Acta Neuropathol 82: 239–259

Braak H, Braak E (1993) Alzheimer neuropathology and limbic circuits. In: Vogt BA, Gabriel M (eds) Neurobiology of cingulate cortex and limbic thalamus. Birkhäuser, Boston, pp 606–626

Braak H, Duyckaerts C, Braak E, Piette F (1993) Neuropathological staging of Alzheimer-related changes correlates with psychometrically assessed intellectual status. In: Corian B, Iqbal K, Nicolini M, Winblad B, Wisniewski H, Zatta PF (eds) Alzheimer's disease: advances in clinical and basic research. Wiley, Chichester, pp 131–137

Braak E, Braak H, Mandelkow EM (1994a) A sequence of cytoskeleton changes related to the formation of neurofibrillary tangles and neuropil threads. Acta Neuropathol 87: 554–567

Braak H, Braak E, Yilmazer D, de Vos R, Janzen E, Bohl J, Jellinger K (1994b) Amygdala pathology in Parkinson's disease, Acta Neuropathol 88: 493–500

Braak H, Braak E, Yilmazer D, de Vos RAI, Jansen ENH, Bohl J, Jellinger K (1995) Nigral and extranigral lesions in Parkinson's disease. J Neural Transm [Suppl 46]: 15–31

Braak H, Braak E, Yilmazer D, de Vos RAI, Jansen ENH, Bohl J (1996) Pattern of brain destruction in Parkinson's and Alzheimer's disease. J Neural Transm 103: 455–490

Dickson DW, Ruan D, Crystal H, Mark MH, Davies P, Kress Y, Yen SH (1991) Hippocampal degeneration differentiates diffuse Lewy body disease (DLBD) from Alzheimer's disease: light and electron microscopic immunocytochemistry of CA2–3 neurites specific to DLBD. Neurology 41: 1402–1409

Fearnley J, Lees A (1994) Pathology of Parkinson's disease. In: Calne DB (ed) Neurodegenerative diseases. Saunders, Philadelphia, pp 545–554

Felleman DJ, van Essen DC (1991) Distributed hierarchical processing in the primate cerebral cortex. Cerebral Cortex 1: 1–47

Gibb WRG, Lees AJ (1991) Anatomy, pigmentation, ventral and dorsal subpopulations of the substantia nigra, and differential cell death in Parkinsons disease. J Neurol Neurosurg Psychiatry 54: 388–396

Gibb WRG, Scott T, Lees AJ (1991) Neuronal inclusions of Parkinson's disease. Mov Disord 6: 2–11

Goldman-Rakic PS, Selemon LD, Schwartz ML (1984) Dual pathways connecting the dorsolateral prefrontal cortex with the hippocampal formation and parahippocampal cortex in rhesus monkey. Neuroscience 12: 719–743

Jellinger K (1990) New developments in the pathology of Parkinson's disease. In: Streifler MB, Korczyn AD, Melamed E, Youdim MBH (eds) Advances in neurology, vol 53. Parkinson's disease: anatomy, pathology, and therapy. Raven Press, New York, pp 1–16

Jellinger K (1991) Pathology of Parkinson's disease. Changes other than the nigrostriatal pathway. Mol Chem Neuropathol 14: 153–197

Jellinger K (1994) Structural basis of dementia in Parkinson's disease. In: Korczyn AD (ed) Dementia in Parkinson's disease. Monduzzi Editore, Bologna, pp 31–38

Jellinger K, Bancher C (1995) Structural basis of mental impairment in Parkinson's disease. Neuropsychiatrie 9: 9–14

Jellinger K, Braak H, Braak E, Fischer P (1991) Alzheimer lesions in the entorhinal region and isocortex in Parkinson's and Alzheimer's diseases. Ann NY Acad Sci 640: 203–209

Kosel KC, van Hoesen GW, Rosene DL (1982) Nonhippocampal cortical projections from the entorhinal cortex in the rat and rhesus monkey. Brain Res 244: 201–213

Lowe J (1994) Lewy bodies. In: Calne DB (ed) Neurodegenerative diseases. Saunders, Philadelphia, pp 51–69

Authors' address: Prof. Dr. med. H. Braak, Department of Anatomy, Theodor Stern Kai 7, D-60590 Frankfurt/Main, Federal Republic of Germany

Morphological substrates of dementia in parkinsonism
A critical update

K. A. Jellinger

Ludwig Boltzmann Institute of Clinical Neurobiology, Vienna, Austria

Summary.   Dementia in parkinsonism is caused by a variety of central nervous system (CNS) lesions, of which the molecular and pathogenic causes are poorly understood but probably include: 1. Degeneration of subcortical ascending systems with neuronal losses in dopaminergic, noradrenergic, serotonergic, cholinergic or multiple systems including the amygdyloid nucleus; 2. limbic and/or cortical Alzheimer and/or Lewy body pathologies, with loss of synapses and neurons, and 3. a combination of these lesions or additional CNS pathologies. In general, degeneration of subcortical neuronal networks appears insufficient to induce severe mental decline although, occasionally, cognitive impairment occurs without apparent cortical lesions. On the other hand, neuritic cortical Alzheimer change showing similar or differential distribution compared to Alzheimer's disease (AD) displays a significant linear correlation with dementia in Parkinsonism. Plaques can be associated with cortical Lewy bodies and, the contribution of each to dementing processes remains unresolved. In a consecutive autopsy series of 610 patients with parkinsonism, the total prevalence of retrospectively assessed dementia was 34.6%. In Parkinson's disease (PD) of the Lewy body type, it was 30.2%, mostly associated with other brain lesions, mainly AD, while only 3.5% of "pure" PD without additional brain pathologies were demented. There was no significant difference in age and duration of illness between demented and non-demented PD patients. Secondary parkinsonian syndromes showed a higher incidence of dementia (56.3%), again with predominant Alzheimer pathology which was present in 73% of the total of demented parkinsonian patients and in almost 82% of the demented PD cases in this series. The specific contribution of cortical and subcortical lesions to mental impairment in parkinsonism, their relationship to AD, and an etiology await further elucidation.

Introduction

In patients with parkinsonism the lifetime risk of developing dementia is about twice as high as in the general population; it averages 10 to 30% increasing to 70% in subjects over 80 years of age (Mayeux et al., 1992;

Marder et al., 1994; Friedman and Barcikowska, 1994; Tison et al., 1995). The motor deficits in parkinsonism mainly result from degeneration of the nigro-striatal system due to loss of nigral neurons, with dopaminergic denervation of the striatum and increased activity of the GABAergic and glutamatergic pathways with inhibition of the corticothalamic "motor loop" (Gerlach and Riederer, 1993; Albin, 1995). However, the nature and molecular basis of mental changes in parkinsonian syndromes remain controversial. Dementia is generally considered to be of subcortical type and, thus, different from that in Alzheimer's disease (AD) (Dubois et al., 1991, 1994; Stern et al., 1993; Tatemichi et al., 1994; Pillon et al., 1994a, b; Troster et al., 1995), although the high incidence of cortical Alzheimer and Lewy body (LB) pathologies in Parkinson's disease (PD), and recent biochemical data, give evidence of severe cortical involvement associated with mental decline in many parkinsonian patients (Boller et al., 1980; Braak and Braak, 1990; Paulus and Jellinger, 1991; Jellinger et al., 1993; Bancher et al., 1993; Vermersch et al., 1993; Jellinger and Bancher, 1995, 1996). Dementia in parkinsonism results from dysfunction of neuronal networks in subcortical and cortical areas re-lated to a variety of brain lesions (Table 1). The present paper focuses on the major morphological correlates of mental decline in parkinsonism based on a large personal autopsy series, and includes morphometric studies of several subcortical nuclei as well as clinico-pathological correlative studies in de-mented and non-demented PD patients.

Material and methods

Between 1957 and 1994, a total of 610 cases with the clinical diagnosis of PD or parkinsonism were neuropathologically examined using conventional, silver and immunohistochemical methods (for tau, ubiquitin, PHF, etc.), and classified according to

Table 1. Brain lesions associated with mental impairment in parkinsonism

1. *Dysfunction of subcortico-cortical neuronal systems*
   a) Degeneration of nigrostriatal dopaminergic pathway $\Rightarrow$ deafferentation of strio-(pre)frontal loops
   b) Cell loss in medial substantia nigra and VTA $\Rightarrow$ mesocorticolimbic dopamine deficiency
   c) Degeneration of noradrenergic systems (locus coeruleus)
   d) Degeneration of serotonergic systems (dorsal raphe nuclei)
   e) Degeneration of cholinergic systems (nucleus basalis of Meynert)
   f) Degeneration of amygdaloid nuclear complex
   g) Combined degeneration of subcortical ascending systems
2. *Cortical pathologies*
   a) Limbic and/or neocortical Alzheimer-type lesions
   b) Cortical Lewy body pathology
   c) Combination of cortical Alzheimer and Lewy body pathologies
3. *Combination of cortical, subcortical and other pathologies*

Table 2. Pathology of Parkinson's disease with dementia (1957–1994)

Neuropathology	Total	With n	dementia %
Id. Parkinson's dis. (IPD)	226	8	3.5
IPD + lacunar state	98	7	7.1
IPD + infarcts	14	1	7.1
IPD + AD/DAT	122	120	98.4
IPD + AT pathology	21	4	19.0
IPD + MIE	11	4	36.4
IPD + MIX (AD/MIE)	7	6	85.7
IPD + other pathology	8	3	37.5
Prim. IPD	507	153	30.2
AD/DAT°	24	24	100.0
Lewy body variant AD	10	10	100.0
MIE/MIX	22	10	45.4
PSP	17	4/1*	23.5
MSA	16	5/2*	31.2
DLBD (without DAT)	1	0	0
CBD, Pick's disease	4	4	100.0
Nigral lesion unclassif.	5	1	20.0
Others, Postenceph. PD	4	0	0
Second. Park. Syndromes	103	58	56.3
Total	610	211	34.6

° With nigral lesion 14; * with AD; *AT* Alzheimer type, but not AD

current diagnostic criteria. AD was diagnosed according to the NIA (Khachaturian, 1995), CERAD criteria (Mirra et al., 1991), and neuritic Alzheimer lesions were staged according to Braak and Braak (1991). The series included 507 cases of PD of Lewy body type and 103 instances of other disorders referred to as secondary parkinsonian syndromes (Table 2). Dementia was assessed retrospectively from hospital records using DSM-III-R (1987) and WHO ICD-9-NA criteria (Berlit, 1987). Comparative studies of psychometrically assessed mental status and staging of Alzheimer pathology were performed prospectively in a cohort of 45 PD patients (mean age 79.4 years) all of Hoehn and Yahr stages IV or V and with a mean duration of illness of 10 years. The severity of dementia was scored by the Mini-Mental State (MMS) (Folstein et al., 1975) not longer than 4 to 6 months prior to death. Neuropathological staging of Alzheimer lesions was performed by semiquantitative assessment of neuritic plaques (NP) and neurofibrillary tangles (NFT) in midfrontal, parietal, superior temporal, entorhinal and hippocampal regions using Bielschowsky's and Reusche's (1991) silver and tau immunostained sections. They were compared with brains of 105 age-matched subjects from the Vienna Prospective Dementia Study using the same methods (for details see Bancher et al., 1996). In 54 cases of PD (mean age 71 years), all Hoehn and Yahr stages IV and V, with or without dementia, neuronal cell counts were performed in defined areas of the medial and lateral substantia nigra zona compacta (SNZC), dorsale raphe nucleus (DRN) and locus coeruleus (LC); controls were 17 age-matched subjects without neuropsychiatric disorders (Paulus and Jellinger, 1991). In another autopsy cohort of 50 PD patients (mean age 76 and 80, respectively), quantitative assessment of neuronal counts and density was performed in the nucleus basalis of Meynert (NBM) and DRN; Alzheimer changes (NP,

NFT) in isocortex and hippocampus were assessed semiquantitatively (see Jellinger, 1987).

Results and discussion

1. Pathology of parkinsonism with dementia

In the consecutive autopsy series of 610 cases with the clinical diagnosis of parkinsonism (PD and other parkinsonian syndromes), the total prevalence of dementia (moderate or severe) was 34.6 %. Since 103 cases or 16.9% displayed other pathological syndromes or secondary parkinsonism, only 507 cases were confirmed as PD of the Lewy body type. The prevalence of dementia in this group was 30.2%; the majority included PD patients with other central nervous system (CNS) pathologies, in particular AD and mixed type (MIX) dementia (AD plus multi-infarct encephalopathy/MIE). Less often PD dementia was associated with MIE alone or other brain lesions including hydrocephalus, recently discussed (Curran and Lang, 1994), or Alzheimer type (AT) pathology not fulfilling the current neuropathological criteria for the diagnosis of AD; only 3.5% of "pure" PD without concurrent brain pathologies suffered from moderat/severe dementia. Additional cerebrovascular lesions (lacunar state, cerebral infarcts) did not increase the inci-

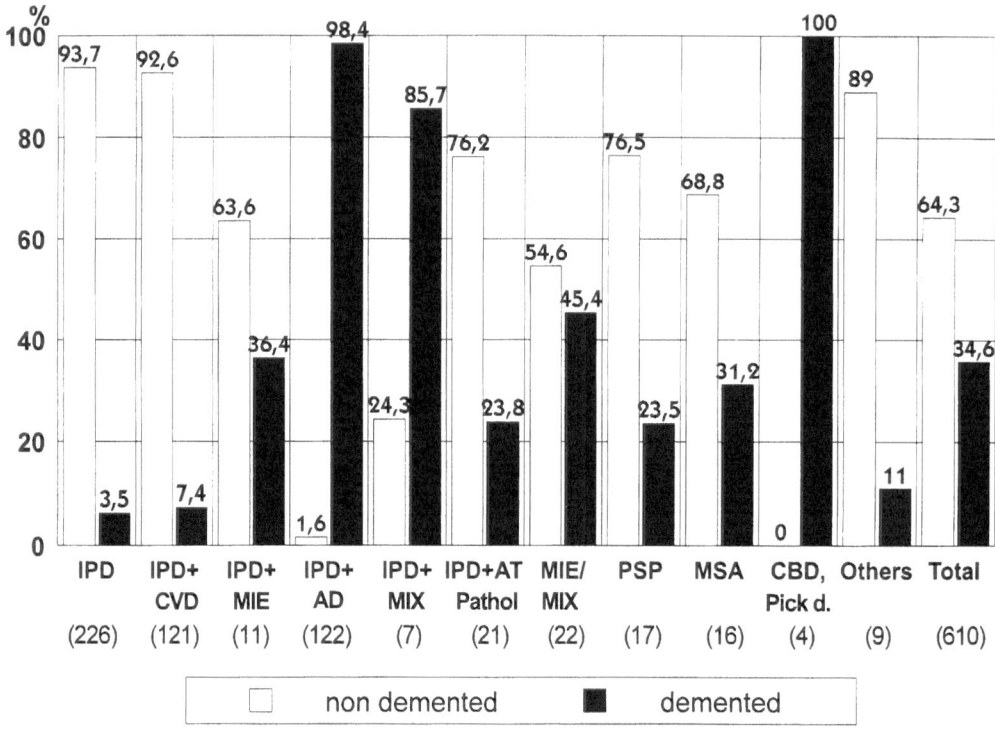

Fig. 1. Incidence of dementia in 610 consecutive autopsy cases clinically presenting with parkinsonian syndromes with different concurrent morphological brain lesions

Table 3. Dementia in Parkinson's disease

Pathology	Non demented			Demented			Total		
	n	Age	Duration	n	Age	Duration	n	Age	Duration
IPD (Lewy body)	218	75.9 ± 4.7	9.1 ± 5.8	8	76.2 ± 6.4	10.9 ± 9.1	226	76.1 ± 5.5	10.0 ± 7.5
IPD + CVD	111	77.1 ± 5.5	5.1 ± 3.0	12	77.6 ± 3.1	5.0 ± 2.1	123	77.3 ± 4.3	5.0 ± 2.3
IPD + AD/ DAT	2	77.0 ± 7.0		120	77.4 ± 7.2	7.2 ± 5.5	122	77.3 ± 7.1	7.2 ± 5.5
IPD + AT pathol.	17	80.0 ± 4.0	7.0 ± 3.0	4	86.0 ± 3.0	4.0 ± 0	21	82.0 ± 4.3	5.0 ± 2.0
IPD + MIX	1	80.0	3.0	6	80.0 ± 2.0	4.2 ± 2.0	7	80.0 ± 2.0	3.5 ± 2.0
IPD + other pathol.	5	67.0 ± 3.0	14.0 ± 9.0	3	78.5 ± 1.5	17.0 ± 13.0	8	70.0 ± 2.4	14.9 ± 10.0
Total	354	76.0 ± 4.3	7.6 ± 5.1	153	79.2 ± 3.8	8.0 ± 5.3	507	77.6 ± 4.4	7.8 ± 5.2

CVD cerebrovascular lesions; *AT* Alzheimer type; *MIX* MIE + AD/DAT

dence of dementia (Table 2; Fig. 1). There was no significant difference in age at death and duration of illness between demented and non-demented PD patients, although those with concurrent AD and cerebrovascular disease (both demented and non-demented) showed shorter duration of disease (Table 3). This finding is in agreement with some other series of PD patients, where age and duration of illness were not significant risk factors for the development of dementia (Caparro-Lefevre et al., 1995). Conversely, the majority of authors report a positive correlation of dementia with either age at onset or present age of PD patients (Dubois et al., 1991; Biggins et al., 1992; Mayeux et al., 1992; Stern et al., 1993; Friedman and Barcikowska, 1994; Troster et al., 1995) and/or shorter survival (Marder et al., 1994; Tison et al., 1995). PD patients with concurrent AD or MIX dementia were not significantly older than those with "pure" PD but showed shorter disease duration, while both demented and non-demented PD subjects with AT pathology but without definite AD showed shorter survival (Table 3).

The incidence of mental impairment was much higher in the group of secondary parkinsonian syndromes (56,3%), with highest incidence in AD patients presenting signs of clinical parkinsonism, in the Lewy body variant of AD, and in rare cases of corticobasal degeneration and Pick's disease. Dementia was less frequent in cases of MIE/MIX and in various neurodegenerative disorders with parkinsonian features, such as multiple system atrophy (MSA) and progressive supranuclear palsy (PSP) (Table 2). The brains of confirmed AD showed mild neuronal loss in SNZC in 14/24 cases (58.3%) which was either diffuse or in dorsolateral regions, usually involved in aging (Fearnley and Lees, 1994); only two of these AD brains showed LB in SN and LC suggesting concurrent subclinical PD. In the group of secondary parkinsonism, the age at death and duration of illness showed considerable variation between the different disorders. AD patients presenting with parkinsonian features were slightly older and had significantly shorter duration than demented PD patients with concurrent AD (Tables 3 and 4).

Table 4. Dementia in secondary parkinsonism

Neuropathology	n	Dementia with without Age/Duration	Total	Age yrs (mean ± S.D.)	Duration
AD/DAT	24		24	80.4 ± 5.6	3.3 ± 2.1
Lewy body variant AD	10		10	74.1 ± 7.1	4.7 ± 2.5
DLBD (without AD)	0		1	44.0	2.0
MIE/MIX	10		11	78.0 ± 5.4	3.0 ± 2.1
PSP	4	71.5/4.5–61.5/11.7	17	65.3 ± 10.3	8.9 ± 7.0
Multisystem atrophy	5	78.5/3.0–59.3/ 5.6	16	60.8 ± 10.6	5.4 ± 3.6
CBD, Pick's disease	4		4	66.0 ± 7.5	6.8 ± 2.5
Nigral lesion unclass.	1		5	78.5 ± 7.2	4.3 ± 2.5
Others, PEP	0		4	70.0 ± 3.0	16.0 ± 4.5
Total	58		103		

Demented subjects with PSP and MSA were significantly older with a shorter clinical course than those non-demented (Table 4). The reason for differences in age of patients and duration of disease between the various types of disorders with and without dementia in our series is unclear.

2. Degeneration of subcortico-cortical systems and dementia

Dopaminergic systems

Degeneration of the nigro-striatal dopaminergic pathway is considered responsible for subtle cognitive changes, such as impaired simultaneous test performance or inefficient reactivation of retrieval process, and other fronto-striatal deficits in PD resulting from disturbances in caudate outflow into the "motor loop" interconnecting the ventral striatum and frontal lobe (Dubois et al., 1991, 1994; Pillon et al., 1994; Rolls, 1994). In PD, there is also considerable damage to the mesocorticolimbic dopamine (DA) system which originates in the ventral tegmental area (VTA) and medial SNZC, both projecting to limbic and prefrontal areas and to the upper brainstem (Nieuwenhuys et al., 1988; Parent and Hazrati, 1995). Neuronal loss in the SNZC in patients with PD, in general, predominantly involves the ventrolateral region leading to dopaminergic denervation particularly of the putamen, which is thought to contribute to the motor symptoms. In demented PD subjects, the medial portion of SNZC shows greater neuronal loss than the lateral regions (Fig. 2; Rinne et al., 1989; Zweig et al., 1993), whereas Duyckaerts et al. (1993) found no differences in SNZC neuronal density between subjects with and without dementia. Dementia in PD patients without concurrent AD has been associated with greater cell depletion in VTA (Zweig et al., 1993), 40–60% loss of DA in limbic and prefrontal cortex (Agid et al., 1990), diffuse reduction of tyrosine hydroxylase (TH) immunoreactivity in the prefrontal cortex (Gaspar et al., 1991), and neocortical monoamine terminal loss (Marie et al., 1995).

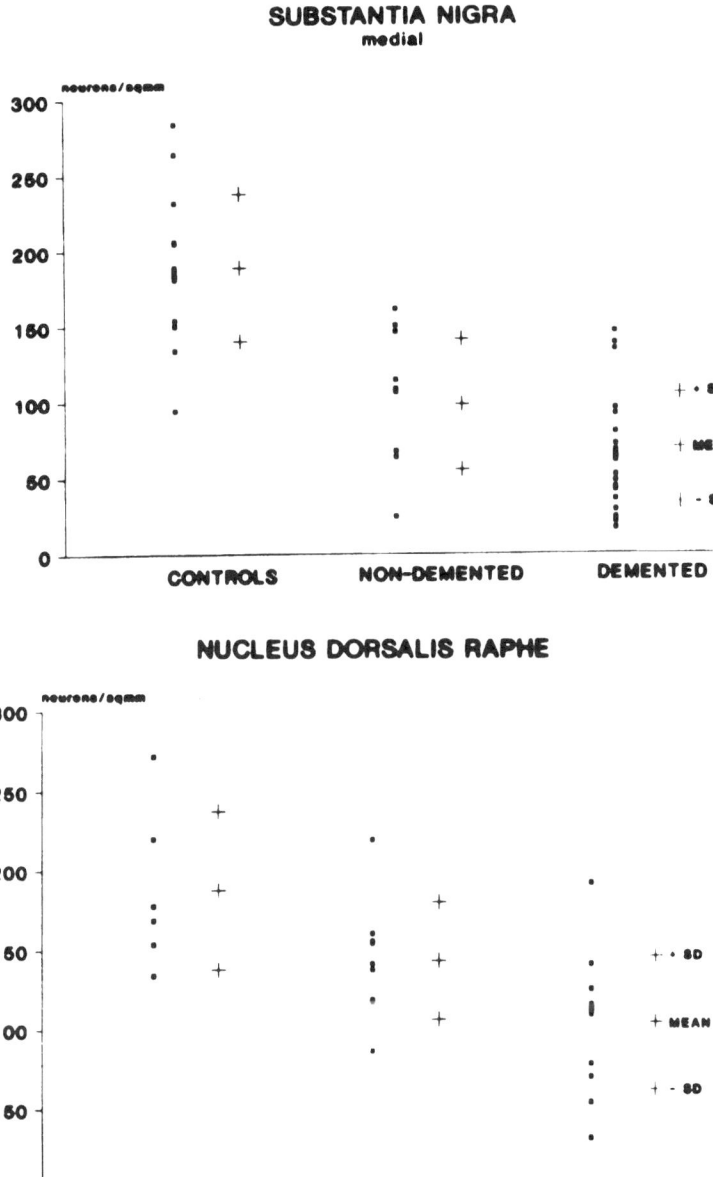

Fig. 2. Neuronal counts per mm² in medial substantia nigra zona compacta, comparing controls, non-demented and demented patients with Parkinson's disease. Symbols represent individual cases. Differences between the groups are significant (p < 0.05)

Modern functional imaging (positron emission tomography — PET) studies show decreased glucose metabolism in frontal areas (Playford et al., 1992; Peppard et al., 1992; Brooks, 1993; Ring et al., 1994). Experimental evidence confirms (Sawaguchi and Goldman-Rakic, 1991) that damage to DA neurons in VTA and medial SNZC and their prefrontal terminals resulting in a mesocorticolimbic DA deficiency is related to cognitive and behavioural dysfunctions in PD (Pillon et al., 1994).

Noradrenergic systems

Degeneration of the noradrenergic system is due to neuronal loss and cell shrinkage in the locus coeruleus (LC) which suffers disease-specific lesions in both PD and AD (German et al., 1992; Zweig et al., 1993; Chan-Palay, 1993; Manaye et al., 1994). In PD, LC cell loss averages 40 to 50% with predominant involvement of the more caudal, compact parts projecting to the cerebellum and spinal cord. Locus coeruleus cell loss is less severe in non-demented PD subjects than in those with depression and/or dementia (Gaspar and Gray, 1984) in whom it approaches the values seen in AD (Chui et al., 1986; Zweig et al., 1993; Chan-Palay, 1993); however other workers describe overlap be-

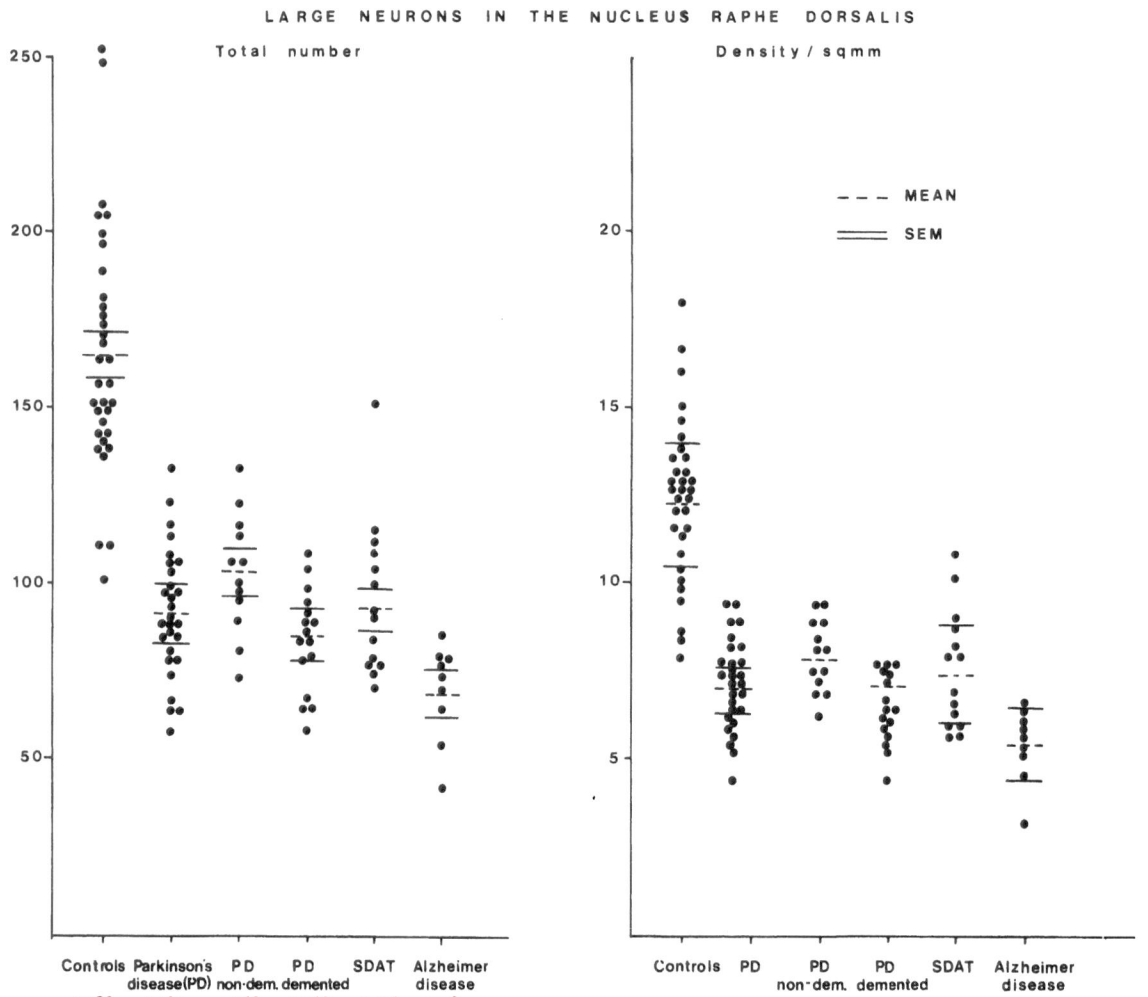

a   b

Fig. 3. Total number (**a**) and mean density (**b**) of large neurons in dorsal raphe nucleus in controls, Parkinson's disease (PD) with and without dementia, Alzheimer's disease (AD) and senile dementia of Alzheimer type (SDAT). Mean values and SEM are indicated by broken and full horizontal lines

tween controls and PD subjects with and without dementia (Duyckaerts et al., 1993). Recent image analyser-assisted morphometric studies of the LC displayed a 39% decrease of large pigmented neurons with a 44% (though not significant) increase of small unpigmented cells in PD patients without depression or dementia, but an 82% loss of large pigmented neurons and 39% decrease of small unpigmented cells in AD (Hoogendijk et al., 1995). Here, neuron depletion in the rostral parts of the LC projecting to the temporal cortex and hippocampus significantly correlates with the density of neuritic AD lesions indicating a retrograde degeneration of subcortical projection nuclei (German et al., 1992) with both cell death and shrinkage due to loss of synapses in their cortical target areas which is an early morphological feature in AD (Heinonen et al., 1995; Masliah, 1995). In contrast, LC in PD brains shows not only cell loss, but also cell shrinkage and a loss of phenotype without relationship to cortical pathology, suggesting primary degeneration of this nucleus (Jellinger, 1991; Hoogendijk et al., 1995). LC damage is most significant in severely demented PD patients without concurrent AD and involves all anatomical levels (Zweig et al., 1993; Chan-Palay, 1993); as a result there is severe deprivation of noradrenergic innervation in the forebrain and neocortex (Javoy-Agid et al., 1989; Gaspar et al., 1991) the effect of which has been related to dementia and depression (Zubenko, 1992) as well as to autonomic dysfunction (Gerlach et al., 1994).

Serotonergic systems

Degeneration of serotonergic systems in PD results from neuronal loss in the dorsal raphe nuclei (DRN) averaging 20 to 40% (Chan-Palay et al., 1992), and is much higher in demented PD patients where it approaches the levels seen in AD (Fig. 3). In PD, the DRN suffers a severe decrease of TH-immunoreactive neurons, whereas the phenylalanine hydroxylase (PH-8) immunostained serotonin-synthesizing neurons are unaffected (Halliday et al., 1992). Many of the remaining neurons contain LB or Lewy neurites (Gai et al., 1995), while in AD, this nucleus has the highest incidence of NFT in brainstem, with more severe involvement than in non-demented aged individuals (Jellinger, 1991). Central serotonergic deficiency that is reflected by reduction of serotonin, its metabolites and receptors in the striatum and medial frontal cortex (Ruberg and Agid, 1988; Agid et al., 1990) has been related to cognitive disorders and depression in patients with PD.

Cholinergic systems

Degeneration of the cholinergic system is characterised by cell loss and shrinkage in the magnocellular posterior parts of the Ch_4 region of the nucleus basalis of Meynert (NBM) projecting to the neocortex (Nieuwenhuys et al., 1988; Steriade and Bisold, 1990), in the pedunculopontine tegmental nucleus (Zweig et al., 1989), the Westphal-Edinger nucleus, and other cholinergic

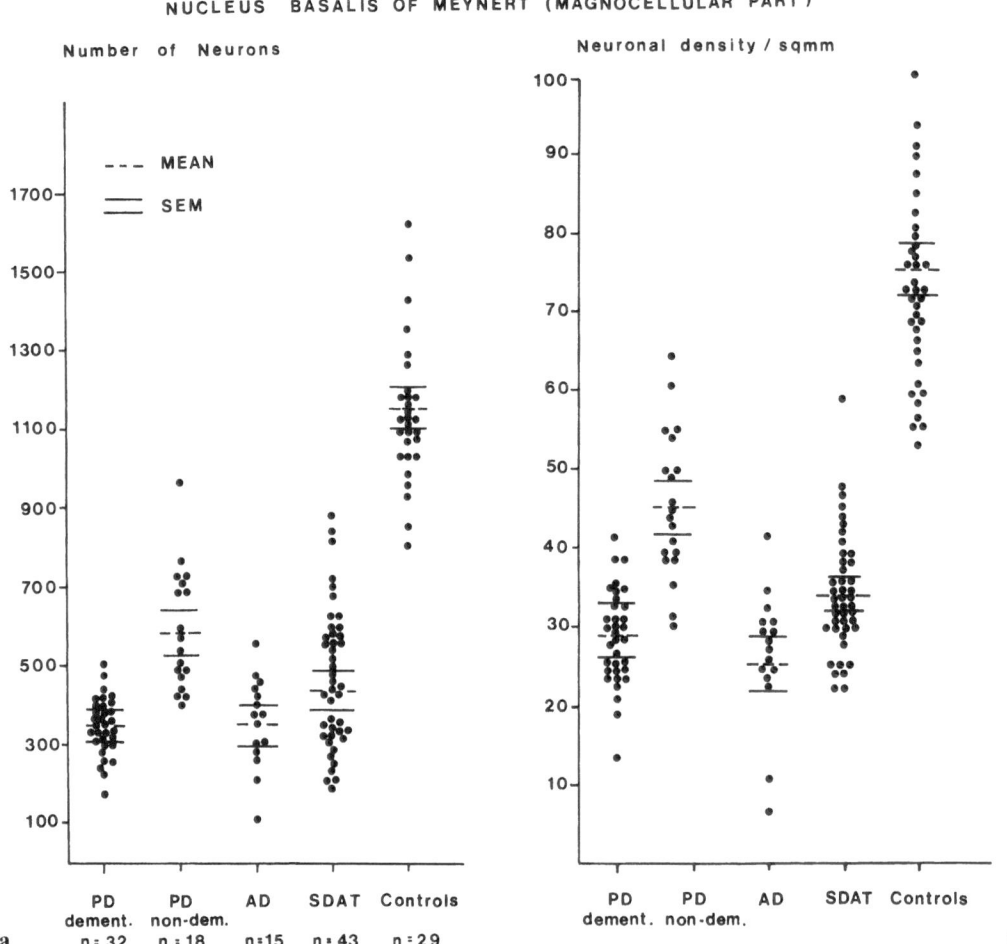

Fig. 4. Total number (**a**) and mean density (**b**) of Ch_4 (magnocellular portion) of nucleus basalis of Meynert in PD patients with and without dementia, AD, DAT, and age-matched controls

brainstem nuclei (see Jellinger, 1991). In PD, NBM cell depletion averages 30 to 40% without correlation to age or duration of illness, and is much higher in demented PD patients (where it approaches the values seen in AD — 50 to 70%) than in PD without dementia (0–40%) who show neuronal losses only slightly higher than normal aged controls (Fig. 4) In other studies, however, considerable NBM cell loss (about 50%) was also seen in non-demented PD subjects (Gaspar and Gray, 1984; Pendleburry and Perl, 1984), with no relationship between cell counts and mental status (Tagliavini et al., 1984; Paulus and Jellinger, 1991). LB in NBM are seen in l00% of cases and NFT in 30 to 65% of all PD brains, whereas neuritic plaques are rare in non-demented cases (Gaspar and Gray, 1984; Jellinger, 1991). Severe depletion of NBM with 75 to 80% loss of large cholinergic neurons is also observed in Lewy body dementia (LBD) (Jellinger and Bancher, 1996). This is accompanied by a decrease of cholinergic innervation of the cortex and hippocampus that may or may not correlate with the severity of NBM cell loss and with mental status

(Gaspar and Gray, 1984; Duyckaerts et al., 1993), although both deficits are usually higher in demented PD (Chui et al., 1986; Ruberg and Agid,1988; Perry et al., 1993). Comparative studies in 50 autopsy cases of PD revealed that in non-demented patients NBM cell loss ranging from 15 to 62% was associated with no or little cortical Alzheimer pathology, while in demented subjects, NBM cell depletion ranged from 64 to 90% and was often, but inconsistently, accompanied by severe cortical AD lesions (Fig. 5). These findings and the demonstration of both frontal cholinergic deficiency and NBM cell loss in PD patients without cognitive deficits (Gaspar and Gray, 1984) suggest that degeneration of the ascending cholinergic system may precede the onset of mental changes and that there is a critical threshold of 75 to 80% neuronal loss and/or shrinkage within the NBM with equivalent cortical cholinergic denervation before dementia becomes apparent (Jellinger, 1987). The loss of cholinergic neurons within the substantia innominata and in the dorso-lateral tegmental nucleus of the brainstem may affect hippocampal and prefrontal structures, either directly or indirectly via the dorsomedial thalamic nucleus (Groenewegen et al., 1993) and may thus contribute to cognitive impairment and subcortical behavioural changes in PD (Reid et al., 1992; Dubois et al., 1994). Although the variable correlations

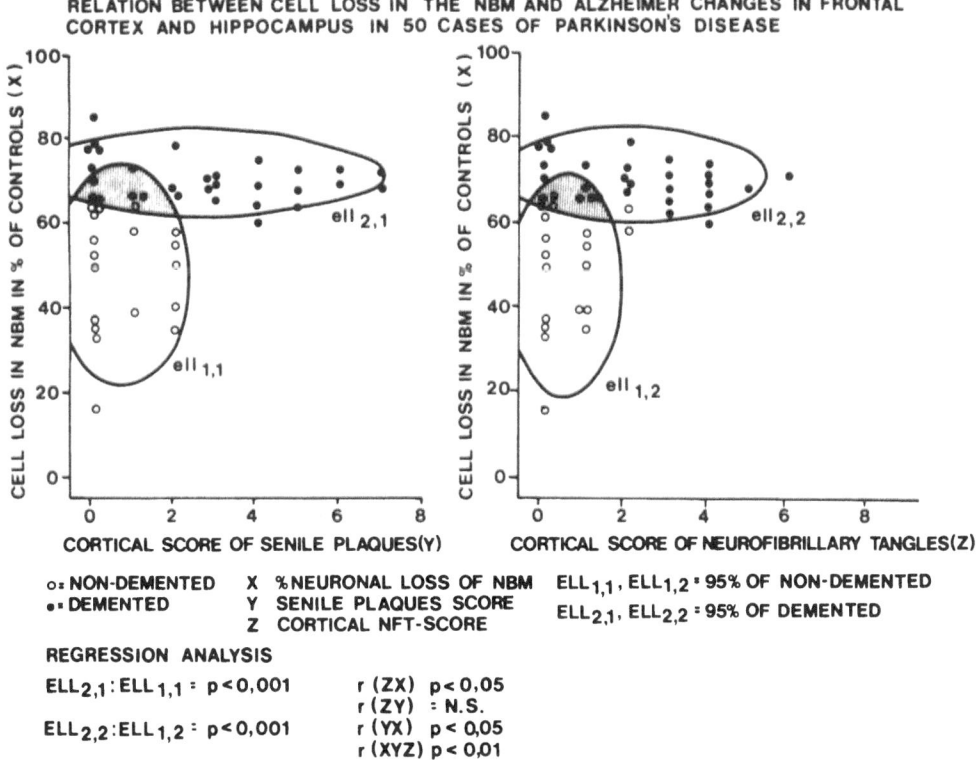

Fig. 5. Relation between neuronal loss in magnocellular portion of nucleus basalis of Meynert (NBM) and neuritic Alzheimer pathology (neuritic plaques, neurofibrillary tangles) in frontal cortex and hippocampus (semiquantitative assessment using 4 degrees of intensity for each region) in 50 autopsy cases of Parkinson's disease with and without dementia

between neuropsychological, morphological and biochemical changes in PD suggest that degeneration of the cholinergic forebrain system often may not be sufficient to induce severe dementia (Duyckaerts et al., 1993), some severely demented PD cases without concurrent AD show considerable degeneration of the NBM with neuronal loss and LB (see Fig. 4). In our current series, there was one woman aged 83 with stage V PD of 28 years' duration and severe dementia (MMS 1/30) who displayed severe degeneration of the NBM with LB as the sole extranigral morphological changes (Table 9). The heterogeneity of degeneration of cholinergic neurons in basal forebrain (Lehericy et al., 1993), and the variability in NBM cell depletion and loss of cholinergic markers in neocortex and hippocampus, irrespective of cortical pathology, suggest a *primary* degenerative process of the cholinergic forebrain system in PD, while secondary retrograde degeneration is proposed for AD (Saper et al., 1986) and has been confirmed by defective retrograde transport of nerve growth factor to NBM in AD (Mufson et al., 1995).

Pathological lesions in the amygdala

This nuclear complex, a primary limbic structure, is anatomically interconnected with the prefrontal cortex, hippocampus, basal forebrain, brainstem and other areas regulating behavioural, mnestic, endocrine and autonomic functions (Sims and Williams, 1992; Amaral et al., 1992; Vereecken et al., 1994; Braak et al., 1994). The amygdala is affected differently in PD and AD. In PD, LB and ubiquitinated Lewy neurites mainly involve the accessory cortical and lateral nuclei, with less severe lesions in the basal and lateral nuclei (Braak et al., 1994). The central accessory nucleus integrates information from the hypothalamus and brainstem to the basal forebrain and centres controlling endocrine or autonomic functions. Since it is involved in "alerting" behaviour and cognition, this damage may contribute to relevant clinical changes in PD. However, since PD-associated amygdala damage has been found to be unrelated to mental impairment, it appears not to be a major cause of dementia. In AD, neuronal loss and shrinkage mainly involve the magnocellular and deep cortical nuclei (Scott et al., 1992; Vereecken et al., 1994) which project to the frontal and temporal isocortex. Destruction of these nuclei may cause "disconnection" of the amygdala from key regions in a similar manner as described for hippocampus in AD (Hof et al., 1994; Braak and Braak, 1994; Samuel et al., 1994a).

In summary, partial lesions of long ascending neuromodulator specific systems in PD due to variable neuronal loss and involvement by LB and ubiquitin-positive degenerating neurites demonstrated in many brainstem nuclei (Gai et al., 1995) may cause demodulation of cognitive programmes suggesting that dementia in PD is associated with pathological involvement of multiple (extra)nigral neuronal populations (Zweig et al., 1993; Pillon et al., 1994). However, the highly variable correlations between neuropsychological, morphological, and biochemical data indicate that degeneration of subcortico-cortical systems is insufficient alone to produce overt dementia in the majority of parkinsonian patients.

3. Cortical pathologies

Besides lesions to subcortico-cortical systems causing bradykinesia and cognitive impairment of "subcortico-frontal" type, considerable cortical pathology is seen in a large proportion of PD patients — particularly in advanced disease stages, and may or may not be associated with severe mental decline. Two major types of cortical pathology are found in demented PD subjects: a) Alzheimer (AD) changes in allocortex and isocortex, and b) cortical Lewy body pathology that may or may not coexist in variable intensity.

Alzheimer type lesions

The morphological features of AD are amyloid deposits (plaques), neuritic changes — NFT, NP and neuropil threads (NPT) — with deposition of paired helical filaments containing hyperphosphorylated tau proteins, and loss of synapses and neurons in the cortex (Terry et al., 1994; Braak and Braak, 1994; Masliah, 1995). Several studies have found strong correlations between dementia scores and neuritic AD changes, in particular neocortical NFT (McKee et al., 1991; Arriagada et al., 1992; Samuel et al., 1994b; Bierer et al., 1995, Bancher et al., 1996). Even more significant correlations are found with dementia and loss of synapse density in isocortex and hippocampus, (Lassmann et al., 1992; Terry et al., 1991, 1994; Samuel et al., 1994b). Both changes have been found to first appear in the entorhinal region (ER) of the hippocampal formation which is thought to be essential for memory and cognition (Hyman et al., 1986; Braak and Braak, 1991; Honer et al., 1992; Bouras et al., 1993; Braak et al., 1994; Samuel et al., 1994a; Heinonen et al., 1995). In human brain, the ER is the main gateway to the hippocampus from numerous isocortical association areas and limbic circuits via the glutamatergic perforant pathway terminating in the hippocampal dentate gyrus, with output modules from interconnected CA1 and subiculum projecting to deep layers of ER and multiple cortical and subcortical areas (Braak and Braak, 1992; Jones, 1993; Samuel et al., 1994a; Insausti et al., 1995). Destruction of the ER by neuritic AD lesions leading to synaptic loss in the dentate and severe involvement of the CA1 and subiculum cause a bi-directional disconnection of the hippocampus from other brain areas which may be responsible for early cognitive deficits in aging, AD and other diseases (Braak et al., 1994; Samuel et al., 1994a). A destructive process related to neuritic AD pathology and loss of synapses causing deafferentation of cortico-cortical and intracortical connections is a major correlate of dementia in AD (Terry et al., 1991, 1994; Lassmann et al., 1993; Hof et al., 1994). Neuritic AD pathology is observed in parkinsonian brain and, in general, is greater in demented subjects than in those with no or only mild mental decline (Braak et al., 1991;

Paulus and Jellinger, 1991; Jellinger et al., 1991; Duyckaerts et al., 1993). It shows similar patterns and hierarchic spreading from allocortex to isocortical areas with early involvement of the (trans)entorhinal region in PD (Jellinger et al., 1991; Arai et al., 1992; Bancher et al., 1993), AD with LB (Hansen et al., 1991), and PSP (Braak et al., 1992). This has been confirmed, at least in part, by immunodetection of the abnormally phosphorylated tau protein triplet, the basic component of AD neurofibrillary degeneration (Vermersch et al., 1993). In non-demented PD patients, it was restricted to the hippocampus, while demented subjects showed higher amounts of tau and NFT in frontal, temporal and entorhinal cortex with severe involvement of the prefrontal area and preservation of the cingulate and occipital cortex, suggesting that frontal lobe dysfunction may contribute to cognitive changes. In parkinsonism-dementia complex of Guam, NFT and a tau triplet similar to that in AD occurs first in limbic areas but is detected in both cortical and subcortical areas thus showing a different distribution compared to AD (Hof et al., 1994; Buée-Scherrer et al., 1995). In PSP with dementia, large amounts of tau doublets are seen in both neocortex and subcortical structures (Vermersch et al., 1994). PNF Tau in LBD is only found in cases with many NFT but not in those with no or only few NFT (Strong et al., 1995). These differences in the regional tau distribution in different neurodegenerative disorders reflect deviations in the distribution of neuritic AD lesions. Although their pattern differs from that in AD with predominant involvement of primary sensory association areas (Vermersch et al., 1992; Hof et al., 1994), these data suggest that neuritic AD pathology does contribute to cognitive disorders in parkinsonian syndromes. Comparative studies of the neuritic AD staging (Braak and Braak, 1991) with the psychometrically assessed mental status in two prospective cohorts of 105 aged individuals (mean age 79.7 years) and 45 age-matched patients with PD displayed similar highly significant linear correlations in both disorders (Fig. 6a,b). This correlation is similar or even better than between numerical counts of neuritic AD lesions or synaptic markers and the psychostatus in earlier studies (Lassmann et al., 1992; Braak et al., 1993; Bancher et al., 1993). In general, demented PD patients show less frequently isocortical neuritic AD lesions, corresponding to stages V or VI (Braak and Braak, 1991) than severely demented aged non-PD subjects with advanced stages of AD. Although the reasons for the differences in intensity and distribution pattern of neuritic AD lesions between PD and AD are unclear, these data suggest that cortical Alzheimer type pathology is a major factor of overt dementia in a substantial proportion of parkinsonian patients. These disorders can be superimposed on cognitive deficits resulting from subcortico-prefrontal dysfunctions related to degeneration of nigral and extranigral and/or other ascending neuronal systems. However, a small proportion of PD patients without dementia present severe cortical AD pathology (Daniel and Lees, 1991), while the substrate of mental impairment in demented PD subjects without evidence of apparent cortical AD lesions (Gaspar and Gray, 1984; Chui et al., 1986; Xuereb et al., 1990; Dubois et al., 1994) also remains enigmatic.

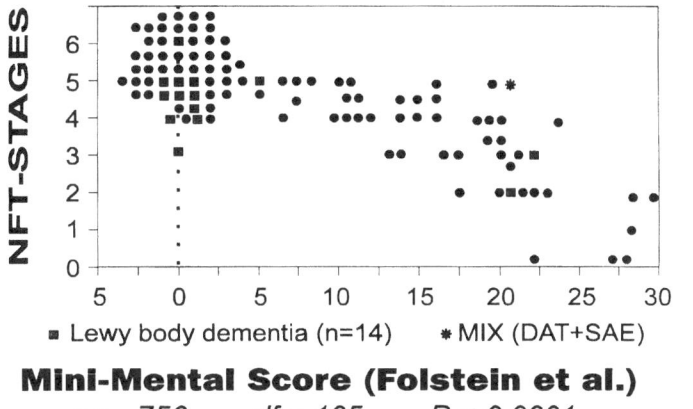

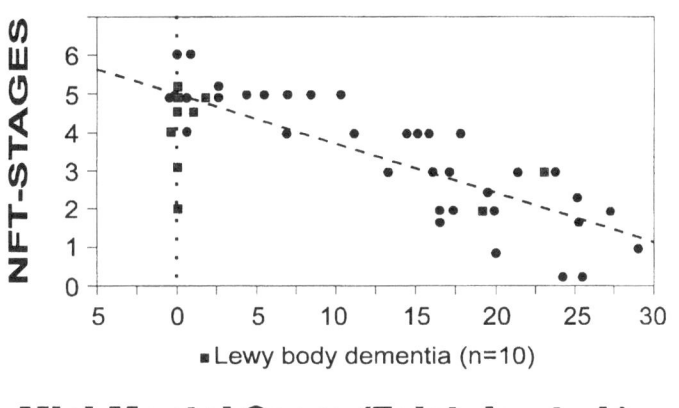

Fig. 6. Correlation of Mini-Mental Score and neuropathological staging of neuritic Alzheimer lesions in 81 consecutive aged individuals (**a**) and in 37 consecutive autopsy cases of Parkinson's disease (**b**)

Cortical Lewy body pathology

Lewy bodies, the anatomical hallmark of PD, are cytoplasmic neuronal inclusions composed of phosphorylated neurofilament proteins undergoing ubiquitination (Pollanen et al., 1993) that involve many subcortical systems and, less frequently, cortical neurons. The presence of cortical LB, seen in almost all PD cases (Hughes et al., 1992; Duyckaerts et al., 1993), has been related to mental decline in a group of conditions referred to as LB dementias or LB disease (LBD) (Kosaka, 1993; Perry et al., 1993; Lippa et al., 1994), senile dementia of LB type (McKeith et al., 1994; Harrington et al., 1994) or

LB variant of AD (Hansen et al., 1993). This progressive dementia with fluctuating confusional states with or without parkinsonian features, the premortem diagnosis of which is difficult (Louis et al., 1995), in some autopsy series constitutes the second most common variant of dementia in the elderly with 7 to 30% (Perry et al., 1993; Harrington et al., 1994b; Ince et al., 1995). In a personal consecutive autopsy series of 290 elderly demented patients (1989–1994), LBD accounted for 14.1% of the total, representing 35.7% of the cases with parkinsonism and dementia (Table 5). Among 29 cases of morphologically confirmed LBD with a mean age of 77.3 ± 10.8 years, seven were clinically diagnosed as PD, seven as AD, 11 as AD + PD, two each as schizophrenia or major depression, one with rigid-akinetic parkinsonism, although almost all patients fulfilled the diagnostic criteria for LBD (McKeith et al., 1994) (Table 6). Morphologically, all brains showed cortical and subcortical LB, 22 had ubiquitin-positive neurites in the CA 2/3 region of hippocampus (Dickson et al., 1994); all except for one brain fulfilled the Khachaturian criteria for AD, but only 14 the CERAD criteria for definite or probable AD; 14 representing AD of plaque-only type (Hansen et al., 1993), while 14 revealed neocortical NFT corresponding to neuritic AD stages V and VI. Twenty brains showed subcortical lesions in SNZC, LC and dorsal vagal nucleus indistinguishable from IPD, while the remaining eight brains had only mild nigral and other subcortical changes associated with AD (Table 7).

The diagnostic criteria of LBD and the question of whether PD with concurrent AD, LBD without neuritic AD pathology, combined LBD and AD or LB variant of AD, and AD with cortical LB represent separate disorders that can sometimes coexist, a spectrum of diseases with pure AD, LBD and PD with or without AD as poles and cases of "mixed pathology" is still under discussion (Lippa et al., 1994; Mann and Snowden et al., 1994). Patients affected by PD or LBD with or without dementia did not demonstrate any differences in association with apolipoprotein E e4 allele (Hardy et al., 1994; Marder et al., 1994; Harrington et al., 1994a; Koller et al., 1995; Lippa et al., 1995; Morris et al., 1996). On the other hand, neurochemically, LBD shares with PD a loss of striatal dopamine and tyrosine hydroxylase (Langlais et al., 1993) and with both AD and PD a decrease in neocortical cholinergic markers related to cell losses in the SNZC and NBM (Perry et al., 1993), while damage to dopamine systems differs between PD and both AD and AD + parkinsonism (Allad et al., 1995; Murray et al., 1995). The relation-

Table 5. Lewy body forms of dementias (autopsies 1989–1994)

Clinical diagnosis	n	LB-dementias	%
Alzheimer disease	211	15	7.1
AD + Parkinsonism	23	6	26.1
Parkinson d. + dementia	56	20	35.7
Dementias, total	290	41	14.1

Table 6. Lewy body dementia (LBD): clinical features. n = 29 (8 males/21 females); age 44–97 (mean 77.3 ± 10.8) years

Clin. DX	PD (n = 7)	AD + PD (n = 11)	AD (n = 7)	Others (n = 4)
Age/death (mean, yrs)	72.9	77.0	78.6	67.8
Duration (mean, yrs)	8.3	6.5	3.7	4.3
PD symptoms	7/7	11/11	3/7	4/4
Fluctuating cognition	5/7	11/11	5/7	4/4
Visual hallucinations	5/7	8/11	3/7	3/4
Repeated falls	7/7	8/11	4/7	3/4
Disturbed consciousness	5/7	10/11	5/7	4/4
Delusions	5/7	10/11	6/7	4/4
Dementia, severe	5/7	11/11	7/7	1/4

Table 7. Lewy body dementia (LBD): neuropathological features

Clin. DX	PD + Dem (n = 7)	AD + PD (n = 11)		AD (n = 7)		Other (n = 4)	
SN lesion	3+	3+	1+	3+	1+	3+	
Cort. LB (>1/mm²)							
Cingulate gyr.	7	6	5	3	4	4	
Parahippocampus	7	6	5	3	4	4	
Lewy neurit. CA2-3	5	5	3	3	3	3	
LB L. coeruleus	7	6	5	3	3	4	
Dors. X n.	7	6	5	3	3	4	
AD pathology							
Khachaturian +	7	6	5	3	4	3	
CERAD/Braak	—	—	—	—	—		
stages 0/1	—	—		—	—	1	A
0/2,3	4			—	—	3	
A/3	—	3		1	—	—	B
B/4	1	1	—	—	1	—	
B/5	2	1	—	—	—	—	C
C/5,6	—	2	4	2	3	—	
Subcort. NFT	—	2	5	2	3	1	

Type of LBD: A) DLBD/no AD (n = 1)
 B) DLBD + AD/plaque type (n = 14) ⎫ LB variant
 C) DLBD + true AD (n = 14) ⎭ of AD

ship between degeneration of SNZC and cortical LB in AD brain (Kazze and Han, 1995) remains to be confirmed but no correlation between nigral degeneration and striatal amyloid plaques in "pure" AD without cortical LB has been observed (Love et al., 1995). Hence, the exact nature of the relationship between PD, LBD, LBD + AD and AD with cortical LB remains debatable, since recent demonstration of biochemical differences between cortical LB in PD and LBD (Smith and Perry, 1994), the different distribution and quantity of neuritic AD pathology, of PHF, and tau proteins, and differences in cholinergic biochemistry argue for a separation of LBD and AD (Perry et al., 1994; Lippa et al., 1994; Harrington et al., 1994b; Strong et al., 1995). Concurrence of AD, PD, LBD and motor neuron disease (Hedera et al., 1995) and similar other neurodegenerative "overlap" syndromes have been reported (Uitti et al., 1995).

Dementia in LBD has been related to either cortical AD or LB lesions, or both, or to dysfunction of subcortical ascending neuronal systems, in particular NBM (Jellinger and Bancher, 1996) and LC, but neither the frequency of cortical LB nor the intensity of neuritic cortical AD pathology were shown to relate directly to the severity of dementia in most LBD patients (Samuel et al., 1996), while a strong correlation with cortical choline acetyltransferase activity and nicotine binding sites suggests a major cholinergic derangement in this disorder (Perry et al., 1993, 1995).

In conclusion, mental decline and dementia in parkinsonian syndromes can be induced by a large variety of brain lesions involving specific cortical and/or ascending subcortico-cortical neuronal loops causing multiple structural and/or functional changes that can be superimposed on the basic degenerative process related to parkinsonian syndromes and which may variously overlap in individual patients. Among our consecutive autopsy series of 610 patients with the clinical diagnosis of PD or other parkinsonian disorders, 73% were PD with or without concurrent CNS pathologies, 55% presenting combination of PD with AD; 26.6% were secondary parkinsonian syndromes, including AD with or without concurrent SN lesions (11%), while PD without

Table 8. Pathology of Parkinson syndromes with dementia
(1957–1994)

Neuropathology	n	%
Id. Parkinson's dis. (IPD)	8	3.8
IPD + cerebrovasc. dis.	12	5.7
IPD + AD/DAT	120	56.9
IPD + AT pathology	4	1.9
IPD + MIX (AD/MIE)	6	2.8
IPD + other pathology	3	1.4
Primary IPD	153	72.5
AD/DAT°	24	11.4
Lewy body variant AD	10	4.7
MIE/MIX	10	4.7
PSP	4/1*	1.9
MSA	5/2*	2.4
CBD, Pick's disease	4	1.9
Nigral lesion unclassif.	1	0.5
Secondary Park. syndromes	58	27.5
Total	211	100.0

°14 with nigral lesion; *with AD

Table 9. Pathology of dementia in Parkinson's disease (1957–1994)

Neuropathology	N	%
IPD (Lewy body)	8	5.2
IPD + lacunar state	7	4.6
IPD + cerebral infarcts	1	0.7
IPD + multiinfarct enceph.	4	2.6
IPD + AD/DAT	120	78.4
IPD + AT pathology	4	2.6
IPD + MIX (AD + MIE)	6	3.9
IPD + degen. nucl. basalis or hydrocephalus	3	2.0
Total	153	100.0

concurrent brain pathologies or with cerebrovascular lesions accounted for 3.8 and 5.7%, respectively. Dementia in other neurodegenerative disorders with parkinsonian features was comparatively rare (Table 8). Within the group of demented PD patients, almost 85% displayed concurrent Alzheimer pathology (including MIX), while only 5.2% showed no additional brain lesions, and 8% additional cerebrovascular lesions (Table 9). These and other data from the literature emphasize the importance of extranigral pathologies for the development of dementia in parkinsonian syndromes. The elucidation of the molecular pathogenic mechanisms and causal interrelationship of these lesions with the basic degenerative disorder causing both parkinsonism and mental decline that may provide a basis for future treatment strategies will be major task of modern neurosciences.

References

Agid Y, Graybiel AM, Ruberg M, Hirsch E, Blin J, Dubois B, Javoy-Agid F (1990) The efficacy of levodopa treatment declines in the course of Parkinson's disease. Do non-dopaminergic lesions play a role? Adv Neurol 53: 83–100

Albin RL (1995) The pathophysiology of chorea/ballism and parkinsonism. Parkinsonism Rel Disord 1: 3–11

Allard PO, Rinner J, Marcusson JO (1995) Dopamine uptake sites in Parkinson's disease and in dementia of the Alzheimer type. Brain Res 637: 262–266

Amaral DG, Price JL, Pitkänen A, Carmichael ST (1992) Anatomical organization of the primate amygdaloid complex. In: Aggleton JP (ed) The amygdala: neurobiological aspects of emotion, memory, and mental dysfunction. Wiley-Liss, New York, pp 1–66

Arai H, Schmidt ML, Lee VM-Y, Hurtig HI, Greenberg BD, Adler CH, Trojanowski JQ (1992) Epitope analysis of senile plaque components in the hippocampus of patients with Parkinson's disease. Neurology 42: 1315–1322

Arriagada PV, Marzloff K, Hyman BT (1992) Distribution of Alzheimer-type pathologic changes in non-demented elderly individuals matches the pattern in Alzheimer's disease. Neurology 42: 1681–1688

Bancher C, Braak H, Fischer P, Jellinger K (1993) Neuropathological staging of Alzheimer lesions and intellectual status in Alzheimer's and Parkinson's disease. Neurosci Lett 162: 179–182

Bancher C, Jellinger K, Lassmann H, et al (1996) Correlations between mental state and quantitative neuropathology in the Vienna Longitudinal Study on Dementia. Eur Arch Psychiatry Clin Neurosci 246: 137–146

Berlit P (1987) Neurologischer Diagnoseschlüssel der internationalen Klassifikation der Krankheiten der WHO (ICD-NA). Springer, Berlin Heidelberg New York Tokyo

Bierer LM, Hof PR, Purohit DP, Carlin L, Schmeidler J, Davis KL, Perl DP (1995) Neocortical neurofibrillary tangles correlate with dementia severity in Alzheimer's disease. Arch Neurol 52: 81–88

Biggins CA, Boyd JL, Harrop FM, Madeley P, Mindham RHS, Randall JI, Spokes EGS (1992) A controlled, longitudinal study of dementia in Parkinson's disease. J Neurol Neurosurg Psychiatry 55: 566–571

Boller F, Mizutani T, Roessmann U, Gambetti P (1980) Parkinson's disease, dementia, and Alzheimer's disease. Clinico-pathological correlations. Ann Neurol 7: 329–335

Bouras C, Hof PR, Morrison JH (1993) Neurofibrillary tangle densities in the hippocampal formation in a non-demented population define subgroups of patients with differential early pathologic changes. Neurosci Lett 153: 131–135

Braak H, Braak E (1990) Cognitive impairment in Parkinson's disease: amyloid plaques, neurofibrillary tangles, and neuropil threads in the cerebral cortex. J Neural Transm [P-D Sect] 2: 45–57

Braak H, Braak E (1991) Neuropathological stageing of Alzheimer-related changes. Acta Neuropathol 82: 239–259

Braak H, Braak E (1992) The human entorhinal cortex: normal morphology and lamina-specific pathology in various diseases. Neurosci Res 15: 6–31

Braak H, Braak E (1994) Pathology of Alzheimer's disease. In: Calne DB (ed) Neurodegenerative disease. Saunders, Philadelphia, pp 585–613

Braak H, Jellinger K, Braak E, Bohl J (1992) Allocortical neurofibrillary changes in progressive supranuclear palsy. Acta Neuropathol 64: 478–483

Braak H, Duyckaerts C, Braak E, Piette F (1993) Neuropathological staging of Alzheimer-related changes correlates with psychometrically assessed intellectual status. In: Corain B, et al (eds) Alzheimer's disease. Advances in clinical and basic research. John Wiley, New York, pp 131–137

Braak H, Braak E, Yilmazer D, De Ros RAI, Jansen ENH, Bohl J, Jellinger K (1994) Amygdala pathology in Parkinson's disease. Acta Neuropathol 88: 493–500

Braak H, Braak E, Mandelkow EM (1994) A sequence of cytoskeleton changes related to the formation of neurofibrillary tangles and neuropil threads. Acta Neuropathol 87: 554–567

Brooks DJ (1993) Functional imaging in relation to parkinsonian syndromes. J Neurol Sci 115: 1–17

Buée-Scherrer V, Buée L, Hof PR, Leveugle-Giolles C, Loerzel AJ, Perl DP, Delacourte A (1995) Neurofibrillary degeneration in amyotrophic lateral sclerosis/parkinsonism dementia complex of Guam: immunochemical characterization of tau proteins. Am J Pathol 68: 924–932

Caporro-Lefevre D, Pecheus N, Petit Y, Duhamel A, Petit H (1995) Which factors predict cognitive decline in Parkinson's disease? J Neurol Neurosurg Psychiatry 58: 51–55

Chan-Palay V (1993) Depression and dementia in Parkinson's disease: catecholaminergic changes in the locus ceruleus. A basis for therapy. Adv Neurol 60: 438–446

Chan-Palay V, Hochli M, Jentsch B, Leonard B, Zetzsche T (1992) Raphe serotonin neurons in the human brain in normal controls and patients with senile dementia of the Alzheimer type and Parkinson's disease. Dementia 3: 253–269

Chui HC, Mortimer JA, Slager U (1986) Pathologic correlates of dementia in Parkinson's disease. Arch Neurol 43: 991–995

Curren T, Lang AE (1994) Parkinsonian syndromes associated with hydrocephalus: case reports, a review of the literature, and pathophysiological hypotheses. Mov Disord 9: 508–520

Daniel SE, Lees AJ (1991) Neuropathological features of Alzheimer's disease in non-demented parkinsonian patients. J Neurol Neurosurg Psychiatry 54: 972–975

Dickson DW, Schmidt ML, Lee VMY, Zhao ML, Yen SH, Trojanowski JQ (1994) Immunoreactivity profile of hippocampal CA2/3 neurites in diffuse Lewy body disease. Acta Neuropathol 87: 269–276

DSM-III-R (1987) Diagnostic and statistical manual of mental disorders, 3rd edn, revised. American Psychiatric Association, Washington

Dubois B, Boller F, Pillon B, Agid Y (1991) Cognitive deficits in Parkinson's disease. In: Boller F, Grafman J (eds) Handbook of neuropsychology, vol 5. Elsevier, Amsterdam, pp 195–240

Dubois B, Malapani C, Verin M, Rogelet P, Deweer B, Pillon B (1994) Cognitive functions and the basal ganglia. The model of Parkinson's disease. Rev Neurol 150: 763–770

Duyckaerts C, Gaspar P, Costa C, Bonnet M, Hauw JJ (1993) Dementia in Parkinson's disease. Morphometric data. Adv Neurol 60: 447–455

Fearnley J, Lees A (1994) Pathology of Parkinson's disease. In: Calne DB (ed) Neurodegenerative diseases. Saunders, Philadelphia, pp 545–554

Folstein MF, Folstein SE, McHugh PR (1975) "Mini-Mental State": a practical method for grading the cognitive state of patients for the clinician. J Psychiatry Res 12: 189–198

Friedman A, Barczikowska M (1994) Dementia in Parkinson's disease. Dementia 5: 12–16

Gai WP, Blessing WW, Blumbergs PC (1995) Ubiquitin-positive degenerating neurites in the brainstem in Parkinson's disease. Brain 118: 1447–1459

Gaspar P, Gray F (1984) Dementia in idiopathic Parkinson's disease. A neuropathological study of 32 cases. Acta Neuropathol 64: 43–52

Gaspar P, Duyckaerts C, Alvarez C, Javoy-Agid F, Boger B (1991) Alterations of dopaminergic and noradrenergic innervations in motor cortex in Parkinson's disease. Ann Neurol 30: 365–374

Gerlach M, Riederer P (1993) The pathophysiological basis of Parkinson's disease. In: Szeleny I (ed) Inhibitors of monoamine oxidase B. Pharmacology and clinical use in neurodegenerative disorders. Birkhäuser, Basle, pp 25–50

Gerlach M, Jellinger K, Riederer P (1994) The possible role of noradrenergic deficits in selected signs of Parkinson's disease. In: Briley M, Marien M (eds) Noradrenergic mechanisms in Parkinson's disease. CRC Press, Boca Raton, pp 59–71

German DC, Manaye KF, White CL (1992) Disease specific patterns of locus ceruleus cell loss. Ann Neurol 32: 667–676

Groenewegen JH, Roeling T, Voorn P, Berendse H (1993) The parallel arrangement of basal ganglia-thalamocortical circuits: a neuronal substrate for the role of dopamine in motor and cognitive functions? In: Wolters EC, Scheltens P (eds) Mental dysfunction in Parkinson's disease. Vrije Universiteit, Amsterdam, pp 3–18

Halliday GM, McCann HL, Pamphlett R, et al (1992) Brain stem serotonin-synthesizing neurons in Alzheimer's disease: a clinico-pathologic correlation. Acta Neuropathol 84: 638–650

Hansen LA, Masliah E, Quidjada-Fawcett S, Rexin D (1991) Entorhinal neurofibrillary tangles in Alzheimer's disease with Lewy bodies. Neurosci Lett 129: 269–272

Hansen LA, Masliah E, Galasko D, Terry RD (1993) Plaque-only Alzheimer disease is usually the Lewy body variant and vice versa. J Neuropathol Exp Neurol 52: 648–654

Hardy J, Crook R, Prihar G, Roberts G, Raghavan R, Perry R (1994) Senile dementia of the Lewy body type has an apolipoprotein E epsilon 4 allele frequency intermediate between controls and Alzheimer's disease. Neurosci Lett 182: 1–2

Harrington CR, Perry RH, Perry EK, Hurt J, McKeith IG, Roth M, Wischik CM (1994a) Senile dementia of Lewy body type and Alzheimer type are biochemically distinct in terms of paired helical filaments and hyperphosphorylated tau protein. Dementia 5: 215–228

Harrington CR, Louwagie J, Rossau R, Vanmechelen E, Perry RH, Perry EK, Xuereb JH, Roth M, Wischik CM (1994b) Influence of apolipoprotein E genotype on senile dementia of the Alzheimer and Lewy body types. Am J Pathol 145: 1472–1484

Hedera P, Lerner AJ, Castellani R, Friedland RP (1995) Concurrence of Alzheimer's disease, Parkinson's disease, diffuse Lewy body disease, and amyotrophic lateral sclerosis. J Neurol Sci 128: 219–224

Heinonen O, Soininen H, Sorvari H, Kosunen O, Paljärvi I, Koivisto E, Riekkienen PJ (1995) Loss of synaptophysin-like immunoreactivity in the hippocampal formation as an early phenomenon in Alzheimer's disease. Neuroscience 64: 375–384

Hoehn MM, Yahr M (1967) Parkinsonism. Onset, progression, and mortality. Neurology 17: 427–442

Hof PR, Morrison JH (1994) The cellular basis of cortical disconnection in Alzheimer disease and related dementing conditions. In: Terry RD, Katzmann R, Bick KL (eds) Alzheimer disease. Raven Press, New York, pp 197–229

Hof PR, Nimchinsky EA, Buée-Scherrer V, Buée L, Nasrallah J, Hottinger AF, Purohit DP, Loerzel AJ, Steele JC, Delacourte A, Bouras C, Morrison JH, Perl DP (1994) Amyotrophic lateral sclerosis/parkinsonism-demetia complex of Guam: quantitative neuropathology, immunohistochemical analysis of neuronal vulnerability, and comparison with related neurodegenerative disorders. Acta Neuropathol 88: 397–404

Hoogendijk WJG, Pall CW, Troost D, Venzweiten E, Swaab DF (1995) Image analysis-assisted morphometry of the locus ceruleus in Alzheimer's disease, Parkinson's disease, and amyotrophic lateral sclerosis. Brain 118: 131–143

Hughes AJ, Daniel SE, Kilford L, Lees AJ (1992) Accuracy of clinical diagnosis of idiopathic Parkinson's disease. A clinico-pathological study of 100 cases. J Neurol Neurosurg Psychiatry 55: 181–184

Hyman BT, Van Hoesen GW, Kromer LJ, Damasio AR (1986) Perforant pathway changes and the memory impairment in Alzheimer disease. Ann Neurol 20: 37–40

Ince PG, McArthur FK, Bjertness E, Torvik A, Condy JM, Edwardson JA (1995) Neuropathological diagnosis in elderly patients in Oslo — Alzheimer's disease, Lewy body disease, vascular lesions. Dementia 6: 162–166

Insausti R, Tunon T, Sobreviela T, Insausti AM, Gonzalo LM (1995) The human entorhinal cortex. A cytoarchitectonic analysis. J Comp Neurol 355: 171–198

Javoy-Agid F, Scatton B, Ruberg M, L'Heureux L, Cervera P, Raisman R, Maloteaus JM, Beck H, Agid Y (1989) Distribution of monoaminergic, cholinergic and gabaergic markers in the human cerebral cortex. Neuroscience 29: 251–269

Jellinger K (1987) Neuropathological substrates of Alzheimer's and Parkinson's disease. J Neural Transm [Suppl] 24: 109–129

Jellinger K (1990) New developments in the pathology of Parkinson's disease. Adv Neurol 53: 1–16

Jellinger K (1991) Pathology of Parkinson's disease: changes other than the nigro-striatal pathway. Mol Chem Neuropathol 14: 153–198

Jellinger K (1994) Structural basis of dementia in Parkinson's disease. In: Korczyn AD (ed) Dementia in Parkinson's disease. Monduzzi Ed, Bologna, pp 31–38

Jellinger KA, Bancher C (1995) Structural basis of mental impairment in Parkinson's disease. Neuropsychiatrie 9: 9–14

Jellinger KA, Bancher C (1996) Dementia with Lewy bodies. Relationship to Parkinson's and Alzheimer's disease. In: Perry E, McKeith JG (eds) Dementia with Lewy bodies. Cambridge University Press, New York, pp 268–286

Jellinger K, Braak H, Braak E, Fischer P (1991) Alzheimer lesions in the entorhinal region and isocortex in Parkinson's and Alzheimer's diseases. Ann NY Acad Sci 640: 203–209

Jellinger KA, Bancher C, Fischer P (1993) Neuropathological correlates of mental dysfunction in Parkinson's disease. In: Wolters ECh, Schelten S (eds) Mental dysfunction in Parkinson's disease. Vrije Universiteit, Amsterdam, pp 141–161

Jones RSG (1993) Entorhinal-hippocampal connections: a speculative view of their function. Trends Neurosci 16: 58–64

Kazee AM, Han LY (1995) Cortical Lewy bodies in Alzheimer's disease. Arch Pathol Lab Med 119: 448–453

Khachaturian ZS (1985) Diagnosis of Alzheimer's disease. Arch Neurol 42: 1097–1105

Klingberg T, Roland PE, Kawashima (1994) The human entorhinal cortex participates in associative memory. Neuro Report 6: 57–60

Koller WC, Glatt SL, Hubble JP (1995) Apolipoprotein E genotypes in Parkinson's disease with and without dementia. Ann Neurol 37: 242–245

Kosaka K (1993) Dementia and neuropathology in Lewy body disease. Adv Neurol 60: 456–463

Langlais PJ, Thal L, Hansen L, Galasko D, Alford M, Masliah E (1993) Neurotransmitters in basal ganglia and cortex of Alzheimer's disease with and without Lewy bodies. Neurology 43: 1927–1934

Lassmann H, Weiler R, Fischer P, Bancher C, Jellinger K, et al (1992) Synaptic pathology in Alzheimer's disease: immunological data for markers of synaptic and large dense core vesicles. Neuroscience 46: 1–8

Lassmann H, Fischer P, Jellinger K (1993) Synaptic pathology of Alzheimer's disease. Ann NY Acad Sci 695: 59–64

Lehericy S, Hirsch EC, Pervera-Plerot P, Javoy-Agid F, Agid Y (1993) Heterogeneity of the degeneration of cholinergic neurons in basal forebrain in patients with Alzheimer's disease. J Comp Neurol 330: 15–31

Lippa CF, Smith TW, Swearer JM (1994) Alzheimer's disease and Lewy body disease: a comparative clinicopathological study. Ann Neurol 35: 81–88

Lippa CF, Smith TW, Saunders AM, et al (1995) Apolipoprotein E genotype and Lewy body disease. Neurology 45: 97–103

Louis ED, Goldman JE, Powers JM, Fahn S (1995) Parkinsonian features of eight pathologically diagnosed cases of diffuse Lewy body disease. Mov Disord 10: 188–194

Love S, Wilcock GK, Matthews SM (1995) No correlation between nigral degeneration and striatal plaques in Alzheimer's disease. Acta Neuropathol 91: 432–435

Manaye KF, Woodward K, McIntire DD, Mann DMA, German DC (1994) Locus coeruleus cell loss on lobar atrophy. Neurodegeneration 3: 205–210

Marder K, Leung D, Tang M, Bell K, Doonelef G, Cote L, Stern Y, Mayeux R (1991) Are demented patients with Parkinson's disease accurately reflected in prevalence surveys? A survival analysis. Neurology 41: 1240–1243

Marder K, Cote L, Tang M, Stern Y, Maestre G, Tycko B, Mayeux R (1994) The risk and predictive factors associated with dementia in Parkinson's disease. In: Korczyn AD (ed) Dementia in Parkinson's disease. Monduzzi Ed, Bologna, pp 51–54

Marie RM, Barre L, Rioux P, Allain P, Lechevalier B, Baron JC (1995) PET imaging of neocortical monoaminergic terminals in Parkinson's disease. J Neural Transm [PD-Sect] 9: 55–71

Masliah E (1995) Mechanisms of synaptic dysfunction in Alzheimer's disease. Histol Histopathol 10: 505–519

Mayeux R, Denaro J, Hemenglido N, Marder K Tang MX, Cote LJ, Stern Y (1992) A population-based investigation of Parkinson's disease with and without dementia. Arch Neurol 49: 492–497

McKeith IG, Fairbairn AF, Bothwell RA, et al (1994) An evaluation of the predictive validity and inter-rater reliability of clinical diagnostic criteria for senile dementia of Lewy body type. Neurology 44: 872–877

Mindham RHS, Biggins CA, Boyd JL, et al (1993) A controlled study of dementia in Parkinson's disease over 54 months. Adv Neurol 60: 469–474

Mirra SS, Heyman A, McKeel D, Sumi SM, Crain BJ, Brownlee LM, Vogel FS, Hughes JP, Van Belle G, Berg L (1991) The Consortium to establish a registry for Alzheimer's disease (CERAD). II. Standardization of the neuropathologic assessment of Alzheimer's disease. Neurology 41: 479–486

Morris CM, Massey HM, Benjamin R, et al (1996) Molecular biology of APO E alleles in Alzheimer's and non-Alzheimer's dementias. J Neural Transm [Suppl] 47: 205–218

Mufson EJ, Conner JM, Kordower JH (1995) Nerve growth factor in Alzheimer's disease. Defective retrograde transport to nucleus basalis. Neuroreport 6: 1063–1066

Murray AM, Weihmueller FB, Marshall JF, Hurtig HI, Gottleib GL, Joyce JN (1995) Damage to dopamine systems differs between Parkinson's disease and Alzheimer's disease with parkinsonism. Ann Neurol 37: 300–312

Nieuwenhuys R, Voogel J, Van Huizen C (1988) The human central nervous system. A synopsis and atlas, 3rd ed. Springer, Berlin Heidelberg New York Tokyo

Parent A, Hazrati L-N (1995) Functional anatomy of the basal ganglia. I. The cortico-basal-ganglia-thalamo-corticol loop. Brain Res Rev 20: 91–127

Paulus W, Jellinger K (1991) The neuropthologic basis of different clinical subgroups of Parkinson's disease. J Neuropathol Exp Neurol 50: 743–755

Pendlebury WW, Perl D (1984) Nucleus basalis of Meynert. Severe loss in Parkinson's disease without dementia (Abstr). Ann Neurol 16: 129

Peppard RF, Martin WRW, Carr GD, Grochoswski E, Schulzer M, et al (1992) Cerebral glucose metabolism in Parkinson's disease with and without dementia. Arch Neurol 49: 1262–1268

Perry EK, Irving D, Kerwin JM, McKeith IG, Thompson P, Collerton D, Fairbairn AP, Ince PG, Morris CM, Cheng AV, Perry RM (1993) Cholinergic transmitter and neurotrophic activities in Lewy body dementia. Similarity to Parkinson's and distinction from Alzheimer disease. Alzheimer Dis Assoc Disord 7: 69–79

Perry EK, Morris CM, Court JA, Choung A, Fairbairn AF, McKeith JG, Irving D, Brown A, Perry RH (1995) Alteration in nicotin binding sites in Parkinson's disease, Lewy body dementia and Alzheimer's disease: possible index of early neuropathology. Neuroscience 64: 385–395

Pillon B, Deweer B, Malapani C, Rogelet P, Agid Y, Dubois B (1994a) Explicit memory disorders of demented parkinsonian patients and underlying neuronal basis. In: Korczyn AD (ed) Dementia in Parkinson's disease. Monduzzi Ed, Bologna, pp 265–271

Pillon B, Michon A, Malapani C, Agid Y, Dubois B (1994b) Are explicit memory disorders of progressive supranuclear palsy related to damage to striatofrontal circuits? Comparison with Alzheimer's, Parkinson's and Huntington's diseases. Neurology 44: 1264–1270

Playford ED, Jenkins IH, Passingham RF, Nutt J, Frackowiak RSJ, Brooks DJ (1992) Impaired mesial frontal and putamen activation in Parkinson's disease: a positron emission tomography study. Ann Neurol 32: 151–161

Reid WJG, Broe, GA, Morris JGL (1992) The role of cholinergic deficiency in neuropsychological deficits in idiopathic Parkinson's disease. Dementia 3: 114–120

Ring HA, Bench CJ, Trimble MR, Brooks DJ, Frakowiak RSJ, Dolan RJ (1994) Depression in Parkinson's disease. A positron emission study. Br J Psychiatry 165: 333–339

Rinne JO, Rummukainen J, Paljärvi J, Rinne UK (1989) Dementia in Parkinson's disease is related to neuronal loss in the medial substantia nigra. Ann Neurol 26: 47–50

Rolls ET (1994) Neurophysiology and cognitive functions of the striatum. Rev Neurol 150: 648–660

Ruberg M, Agid Y (1988) Dementia in Parkinson's disease. In: Iversen LL, Iversen SD, Snyder SH (eds) Handbook of psychopharmacology, vol 20. Psychopharmacology of aging nervous system. Plenum Press, New York, pp 157–206

Samuel W, Masliah E, Hikll R, Butters N, Terry R (1994a) Hippocampal connectivity and Alzheimer's dementia: effects of synapse loss and tangle frequency in a two-component model. Neurology 44: 2081–2088

Samuel W, Terry RD, DeTeresa R, et al (1994b) Clinical correlates of cortical and nucleus basalis pathology in Alzheimer dementia. Arch Neurol 51: 772–778

Samuel W, Galasko D, Masliah E, Hansen LA (1996) Neocortical Lewy body counts correlate with dementia in the Lewy body variant of Alzheimer's disease. J Neuropathol Exp Neurol 55: 44–52

Saper CD, German DC, White CL (1985) Neuronal pathology in the nucleus basalis of Meynert and associated cell groups in senile dementia of the Alzheimer's type. Possible role of cell loss. Neurology 35: 1089–1095

Sawaguchi T, Goldman-Rakic PS (1991) D1 dopamine receptors in prefrontal cortex: involvement in working memory. Science 251: 247–250

Scott SA, DeKosky ST, Sparks DL, Knox CA, Scheff SW (1992) Amygdala cell loss and atrophy in Alzheimer's disease. Ann Neurol 32: 555–563

Sims KS, Williams RS (1990) The human amygdaloid complex. Neuroscience 36: 449–472

Smith MA, Perry G (1994) Is a Lewy body always a Lewy body? In: Korczyn AD (ed) Dementia in Parkinson's disease. Monduzzi Ed, Bologna, pp 187–193

Steriade M, Biesold D (1990) Brain cholinergic systems. Oxford University Press

Stern Y, Marder K, Tang MX, Mayeux R (1993) Antecedent clinical features associated with dementia in Parkinson's disease. Neurology 43: 1690–1692

Stern Y, Richards M, Sano M, Mayeux R (1993) Comparison of cognitive changes in patients with Alzheimer's and Parkinson's disease. Arch Neurol 50: 1040–1045

Strong C, Anderton BH, Perry RH, et al (1995) Abnormally phosphorylated tau protein in senile dementia of Lewy body type and Alzheimer disease; evidenced that the disorders are distinct. Alzheimer Dis Assoc Disord 9: 215–222

Tagliavini F, Pilleri G, Bouras C, Constantinidis J (1984) The basal nucleus of Meynert in idiopathic Parkinson's disease. Acta Neurol Scand 69: 20–28

Tatemichi TK, Sacktor N, Mayeux R (1994) Dementia associated with cerebrovascular disease, other degenerative diseases, and metabolic disorders. In: Terry RD, Katzman R, Bick KL (eds) Alzheimer disease. Raven Press, New York, pp 123–166

Terry RD, Masliah E, Salmon DP (1991) Physical basis of cognitive alterations in Alzheimer's disease: synapse loss is the major correlate of cognitive impairment. Ann Neurol 30: 572–580

Terry RD, Masliah E, Hansen LA (1994) Structural basis of the cognitive alterations in Alzheimer's disease. In: Terry RD, Katzman R, Bick KL (eds) Alzheimer disease. Raven Press, New York, pp 179–196

Tison F, Dartigues JF, Auriacombe S, Letenneur L, Boller F, Alperovich A (1995) Dementia in Parkinson's disease. A population-based study in ambulatory and institutionalized individuals. Neurology 45: 705–708

Troster AL, Paolo AM, Lyons KE, Glatt SL, Hubble JP, Koller WC (1995) The influence of depression on cognition in Parkinson's disease. A pattern impairment distinguishable from Alzheimer's disease. Neurology 45: 672–676

Uitti RJ, Berry K, Yasuhara O, et al (1995) Neurodegenerative "overlap" syndrome: clinical and pathological features of Parkinson's disease, motor neuron disease, and Alzheimer's disease. Park Rel Dis 1: 21–34

Vereecken ThHLG, Vogels OJM, Nieuwenhuys R (1994) Neuron loss and shrinkage in the amygdala in Alzheimer's disease. Neurobiol Aging 15: 45–54

Vermersch P, Frigard B, Delacourte A (1992) Mapping of neurofibrillary degeneration in Alzheimer's disease; evaluation of heterogeneity using the quantification of abnormal tau proteins. Acta Neuropathol 85: 48–54

Vermersch P, Delacourte A, Javoy-Agid F, Hauw JJ, Agid Y (1993) Dementia in Parkinson's disease: biochemical evidence for cortical involvement using the immunodetection of abnormal tau proteins. Ann Neurol 33: 445–450

Vermersch P, Robitaille Y, Bernier L, et al (1994) Biochemical mapping of neurofibrillary degeneration in a case of progressive supranuclear palsy: evidence for general cortical involvement. Acta Neuropathol 87: 572–577

Xuereb JH, Tomlinson BE, Irfing D, et al (1990) Cortical and subcortical pathology in Parkinson's disease: relationship to parkinsonian dementia. Adv Neurol 53: 35–40

Zubenko GS (1992) Biological correlates of clinical heterogeneity in primary dementia. Neuropsychopharmacology 6: 72–93

Zweig RM, Cardilio JE, Cohen M, Giere S, Hedreen JC (1993) The locus ceruleus and dementia in Parkinson's disease. Neurology 43: 986–991

Author's address: Prof. Dr. K. A. Jellinger, Ludwig Boltzmann Institute of Clinical Neurobiology, Wolkersbergenstrasse 1, A-1130 Vienna, Austria

The Lewy body variant of Alzheimer disease

L. A. Hansen

Departments of Pathology and of Neurosciences, School of Medicine, University of
California, San Diego, CA, U.S.A.

Summary. The Lewy body variant of Alzheimer disease (LBV) occupies a
messy middle ground between Alzheimer disease (AD) on the one hand, and
pure Lewy body diseases (Parkinson's disease or diffuse Lewy body disease),
on the other. In addition to brainstem and neocortical Lewy bodies, LBV
brains have enough neocortical neuritic plaques to meet diagnostic criteria for
AD. However, neurofibrillary pathology in LBV is modest, since tangle densities in LBV are typically intermediate between AD and age-matched controls or pure Lewy body disease brains. Apolipoprotein E-4 is overrepre-
sented in LBV, as it is in AD but is not in PD or diffuse Lewy body disease
(DLBD). Neurologically, LBV patients often display sufficient parkinsonian
signs to separate them from AD, but these findings are usually too subtle to
warrant clinical diagnoses of Parkinson's disease (PD). Neuropsychological
deficits in LBV include a subcortical dementia pattern similar to DLBD, and
more severe global cognitive impairment reminiscent of AD.

Introduction

The Lewy body variant of Alzheimer disease (LBV) is a clinical and
pathologic entity (Hansen, 1990). LBV patients present during life with
dementia, often with mild parkinsonian signs. At autopsy, LBV brains show
subcortical and neocortical Lewy bodies, accompanied by sufficient numbers
of neocortical senile plaques (both diffuse and neuritic) to meet widely accepted neuropathological criteria for Alzheimer disease (AD) (Khachaturian,
1985; Mirra, 1993).

The nomenclature of dementia associated with Lewy bodies is unsettled.
Many cases which we designate as LBV would be labelled by other investigators as combined AD and PD (Ditter, 1987), AD with PD-related changes
(Mirra, 1993), AD with incidental Lewy bodies (Joachim, 1988), AD with
concomitant Lewy body disease (Hansen, 1989), senile dementia of the Lewy
body type (Perry, 1990), diffuse Lewy body disease (DLBD), common form
with plaques and/or tangles (Kosaka, 1990), or DLBD. We reserve the designation DLBD for the much smaller percentage of dementia brains with
brainstem and neocortical Lewy bodies lacking accompanying neocortical

AD pathology. Other investigators, however, make no such distinction (Dickson, 1987).

This nosologic tower of Babel results from differing underlying assumptions about whether LBV is primarily a form of AD with Lewy body flavouring, or fundamentally a type of Lewy body disease contaminated by non-specific age appropriate AD pathology. Such an issue is difficult to resolve, since most clinical, neuropsychological, and neuropathologic parameters place LBV squarely in-between AD and pure PD, or DLBD. This investigation describes the extent of AD neuropathology in LBV compared to pure DLBD and age-matched controls. We also review the clinical and neuropsychological characteristics of LBV patients compared to those with pure AD.

Materials and methods

Subjects

The neuropathologically characterized brains in this study came from patients prospectively evaluated at a number of sites associated with the Alzheimer's Disease Research Center at the University of California, San Diego, over the past 10 years. Most of these patients carried the clinical diagnosis of probable or possible AD, and none were clinically diagnosed as Parkinson's disease, although some had mild parkinsonian features. Non-AD, non-LBV, and non-DLBD control brains utilized for comparative purposes came from two groups. There were a small number of carefully examined cognitively intact elderly controls. A second group which we term "pathological controls", came from clinically demented patients who at autopsy demonstrated non-AD, non-LBV and non-DLBD causes for dementia (e.g., multi-infarct dementia, progressive supranuclear palsy, or Pick's disease).

Specimen processing

At autopsies, brains were removed in the usual fashion within 24 hours of death and divided sagittally while fresh. The left hemibrains were fixed in 10% buffered formalin while the right hemibrains were frozen at $-70°C$ for chemical analysis. Following 10–14 days of fixation, the formalin fixed left hemibrains were weighed, examined externally, and sectioned. The calculated fixed whole brain weights were obtained by doubling the hemibrain values. Tissue blocks were taken from midfrontal cortex, superior temporal gyrus, inferior parietal cortex, anterior and posterior levels of the hippocampus, basal ganglia, substantia innominata, mesencephalon, and rostral pons. H&E stained sections 6–8 μm thick were prepared from these blocks. Sections 10 μm thick were stained with thioflavin-S and viewed with ultraviolet illumination and 440 μm bandpass wavelength excitation filters. After each neocortical section was surveyed to find areas with the most lesions, three X125 fields were counted for senile plaques, and three X500 fields for neurofibrillary tangles. The results were then averaged to provide single plaque and tangle counts for each neocortical region from every case. Neocortical senile plaque counts were subdivided into total and neuritic. Total plaques included both diffuse and neuritic varieties, and these counts were used for assigning cases to the AD and LBV categories based on National Institute on Aging (NIA), criteria (Khachaturian, 1985) which do not specify plaque type. Neuritic plaque frequencies were used for consigning cases to the AD category if the corresponding age-adjusted plaque score in the Consortium to Establish a Registry for Alzheimer Disease (CERAD) diagnostic scheme indicated either "definite AD" or "probable AD" (Mirra, 1993). The LBV was defined as the combination of AD, diagnosed according to NIA and CERAD criteria, with subcortical and neocortical Lewy bodies in brains from patients presenting with dementia, not Parkinson's disease. All LBV had to have Lewy bodies in the locus ceruleus, substantia nigra, or substantia innominata as well as neocortex. The presence of neocortical Lewy bodies in a brain with AD qualified the case as a LBV, and no distinctions were made between cases with sparse or plentiful Lewy bodies. If the brain of a clinically demented patient showed only brainstem and neocortical Lewy bodies and lacked significant neocortical senile plaque pathology, it was designated diffuse Lewy body disease (DLBD). Neocortical quantification's of choline acetyltransferase and synaptophysin were performed on the frozen right hemibrains as previously reported (Hansen, 1990; Alford, 1994).

AD neuropathology staging

In order to assess the extent of AD neurofibrillary pathology in our AD, LBV, DLBD, and non-AD control brains, we employed a modification of the neuropathological staging of Alzheimer-related changes scheme described by Braak and Braak (1991). Our modification of the Braak and Braak staging protocol is modelled closely on the original, and attempts to parallel its six developmental stages of neurofibrillary pathology in AD; stages I and II, the transentorhinal, stages III and IV, the limbic, and stages V and VI, the isocortical. In our modification, neurofibrillary pathology is quantified in layer two of the entorhinal cortex at the level of the mammillary bodies, where neuronal clusters give rise to the perforant path. Using thioflavin-S preparations, we counted tangles in at least five neuron clusters in lamina two, and averaged the results to give each brain an entorhinal cortex layer two neuron cluster tangle count. We also assessed neurofibrillary tangles in the midfrontal, inferior parietal, and superior temporal gyrus for assigning cases to stages V or VI in those instances where neurofibrillary pathology advanced beyond the confines of the medial temporal lobe. In stage I, entorhinal cortex layer two neuron clusters contain an average of 0 tangles, but there are scattered tangles in the adjacent transentorhinal cortex. In stage II, entorhinal layer two tangle counts average near 7, in stage III they are around 11, in stage IV the mean counts are 15, and in stages V or VI entorhinal layer two neuron tangle counts typically exceeded 20. Distinctions between stages V and VI are based on the frequency of neocortical tangles in the frontal, temporal, and parietal lobes.

Results

Comparisons between neuropathologic parameters in AD and LBV are displayed in Table 1. The AD and LBV patients did not differ in age at death, degree of dementia, total plaque counts (diffuse plus neuritic) in neocortex and hippocampus, or in midfrontal cortex synaptophysin levels. The LBV patients had shorter disease durations, and brains were less atrophic, as measured by brain weight. LBV brains had fewer neuritic plaques and neurofibrillary tangles in neocortex and hippocampus. Unlike the preceding parameters, all of which indicate greater pathologic abnormality in AD than LBV, neocortical choline acetyltransferase loss was greater in LBV than AD.

Table 1. Neuropathological parameters: AD vs. LBV

	AD	N	LBV	N	P
Age	*78.3*	*255*	*77.7*	*44*	*NS*
Blessed score	26	123	24.5	26	NS
Duration	8.9	128	6.8	24	0.02
Brain weight	1,113	238	1,167	40	0.05
MF-total plaques	37.9	217	35.8	32	NS
IP-total plaques	38.2	216	34.6	32	NS
ST-total plaques	34.4	219	30.3	32	NS
HIPPO-total plaques	10.7	215	8.5	33	NS
MF-neuritic plaques	27.5	81	18.6	23	0.02
IP-neuritic plaques	29.5	82	21.3	23	0.02
ST-neuritic plaques	24.2	70	16.7	23	0.05
HIPPO-neuritic plaques	9.4	69	5.4	22	0.001
MF-tangles	2.9	219	0.5	32	0.001
IP-tangles	4.3	217	0.9	32	0.001
ST-tangles	5.3	220	1.8	32	0.001
HIPPO-tangles	9.6	212	3.0	33	0.001
MF-synaptophysin	73.8	44	78.8	17	NS
MF-ChAT	140.3	97	77.0	15	0.05
IP-ChAT	112.8	96	53.3	15	0.02

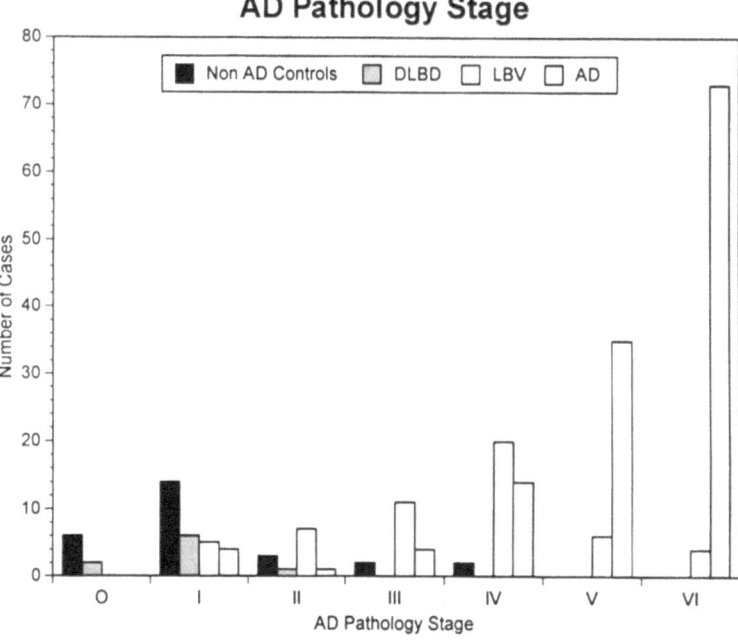

Fig. 1. Most brains with pure Alzheimer disease (AD) were in neurofibrillary pathology stages V and VI. Lewy body variant (LBV) brains were distributed in a bell curve around an average stage III-IV. Non-AD control brains and pure diffuse Lewy body disease (DLBD) brains were essentially confined to stages 0–II

Table 2. Neuropathological paramenters: LBV vs. controls

	LBV	N	Controls	N	P
Age	*77.7*	*44*	*79.8*	*36*	*NS*
Blessed score	24.5	26	6.7	11	0.001
Brain weight	1,167.2	40	1,161.6	29	NS
MF-total plaques	35.8	32	2.4	29	0.001
IP-total plaques	34.6	32	2.0	25	0.001
ST-total plaques	30.3	32	1.9	27	0.001
HIPPO-total plaques	8.5	33	0.1	27	0.001
MF-tangles	0.5	32	0	29	0.01
IP-tangles	0.9	32	0	25	0.001
ST-tangles	1.8	32	0	27	0.001
HIPPO-tangles	3.0	33	1.5	27	NS
MF-ChAT	77.0	15	336.3	14	0.001
IP-ChAT	53.3	15	346.7	13	0.001
ST-ChAT	130.5	15	351.9	14	0.002
HIPPO-ChAT	362.4	15	750.6	5	NS

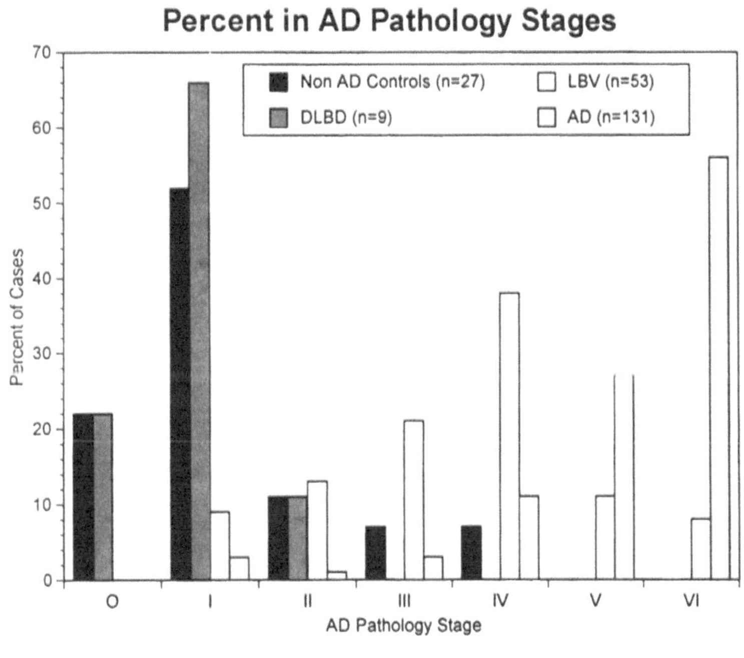

Fig. 2. The same differences in neurofibrillary pathology stages are better seen when expressed as percentages of cases in each stage from each comparative category. *AD* Alzheimer disease, *LBV* Lewy body variant of Alzheimer disease, *DLBD* diffuse Lewy body disease

Table 2 compares the same neurologic parameters in LBV versus age-matched elderly cognitively intact controls. Compared to these controls, the LBV's had more total plaques (diffuse plus neuritic) in neocortex and hippocampus, more tangles in neocortex, but not hippocampus, and an extensive loss of neocortical choline acetyltransferase.

Figures 1 and 2 display the results of Alzheimer disease pathological staging for AD, LBV, DLBD, and non-AD controls. Figure 1 shows the total number of cases at each pathological stage in each category, while Fig. 2 displays the percentages of those totals at each developmental stage for each of the four categories of patients. More than 80% of the AD brains were in pathology stages V and VI, indicating that these brains had numerous neurofibrillary tangles both in entorhinal cortex and neocortex. This is approximately equivalent to the designation of "plaque and tangle" AD, sometimes juxtaposed to the concept of "plaque only or plaque predominate" AD (Hansen, 1993), which corresponds to stages less than V. Over 80% of the non-AD elderly controls, and all the DLBD brains were at stage 0, I, or II. The LBV were distributed in a bell curve over the middle neuropathologic stages, with a mean AD neurofibrillary pathology between stages III and IV.

Discussion

AD pathology in LBV

The neuropathological lesions of AD are diffuse plaques, neuritic plaques, and neurofibrillary tangles. Tables 1 and 2 and Figs. 3 and 4 demonstrate that LBV brains almost invariably have more of these lesions than age-matched controls or pure DLBD, but fewer than are encountered in pure AD. These results confirm and expand upon those previously reported.

Many investigators have shown that AD with Lewy bodies has fewer neocortical tangles than AD without Lewy bodies, despite comparable num-

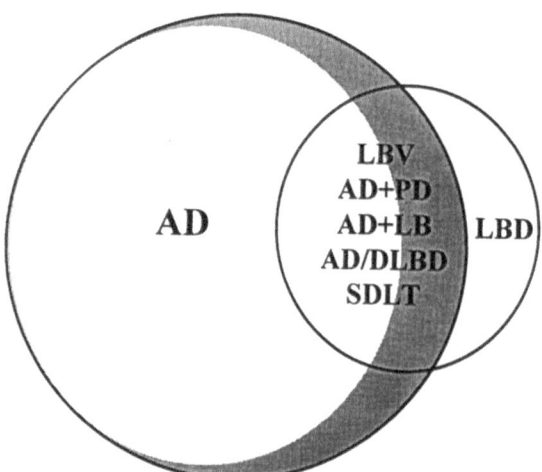

Fig. 3. The most nosologically contentious category in dementia associated with Lewy bodies is the combination of Lewy body disease (LBD) and neocortical "plaque only or plaque predominant" Alzheimer disease (AD), represented here by the gray-shaded crescent. *LBV* Lewy body variant, *PD* Parkinson's disease, *LB* Lewy body, *DLBD* diffuse LB disease, *SDLT* senile dementia of the LB type

bers of senile plaques in the two conditions (Bergeron, 1990; Bierer, 1990;
Hansen, 1990, 1993). These studies have not often subdivided plaque types
into diffuse and neuritic varieties, however. Our results (Table 1) show that
while LBV's have the same number of total senile plaques as AD, and far
more than age-matched controls (Table 2), a higher percentage of the senile
plaques in LBV are diffuse rather than neuritic. Since many lines of evidence
indicate that diffuse plaques precede and progress into neuritic ones, this
observation is consistent with the hypothesis that AD pathology in LBV is in
an earlier developmental stage than that encountered in AD brains lacking
Lewy bodies. This speculation is further supported by the results of
neuropathologic staging in AD, LBV, and controls, as discussed below.

The comparative paucity of neocortical tangles in LBV caused us to re- re-
evaluate the phenomenon of neocortical "plaque only or plaque predominant" AD, which we had reported in 1987 as constituting up to 30% of AD
among patients over age 74 (Terry, 1987). We found that 75% of all neocortical "plaque only or plaque predominant" AD is LBV and, conversely, that
two-thirds of LBV brains have "plaque only or plaque predominant" neocortical AD pathology (Hansen, 1993). In the neuropathological evolution of
AD, the transition from neocortical "plaque only or plaque predominate"
pathology to "plaque and tangle" neocortical AD pathology occurs at stage
IV (Braak, 1991).

The average AD neurofibrillary pathology stage achieved by most LBV
brains (stage III or IV) is intermediate between that seen in elderly non-AD
controls (stage I), and the stage occupied by most brains with only AD
pathology (stage V or VI). (see Figs. 1 and 2). This result could be predicted
from earlier studies. In 1991, we reported that LBV brains have more
entorhinal layer two neurofibrillary tangles than controls (p < 0.001), but
fewer than pure AD brains (p < 0.02) (Hansen, 1991). The results of AD
staging predictably confirm this finding in the present larger series, since our
staging protocol hinges on entorhinal neurofibrillary pathology. Also in 1991,
Ince et al. similarly found that brains which they designated as senile dementia
of the Lewy body type (SDLT) had more temporal lobe archicortical Alzheimer-type pathology (tangles, senile plaques, and granulovacuolar
degeneration) than did age-matched controls or PD patients. Again, the
Alzheimer pathology was less severe in their Lewy body cases than in pure
AD.

Apolipoprotein E allele ε-4 in LBV

Apolipoprotein E ε-4 (APOE-4) allelic frequency is increased in AD compared to controls (Strittmatter, 1993), but not in PD (Nalbantoglu, 1994). We
reasoned therefore that measuring APOE-4 allelic frequency in LBV would
help determine whether LBV was genotypically more closely related to AD
or to PD. We found an increased APOE-4 allelic frequency in LBV of 29%
(Glasko, 1994), compared to a control frequency of 14.2% in a previous large
study by Menzel (1983) (p < 0.001 for LBV versus controls). This increased

frequency in LBV did not differ significantly from the increased frequency of 39.6% we measured in AD (AD compared to LBV, p = 0.12, chi-square). An avalanche of simultaneous publications confirms this finding, although comparisons of the reported data are complicated by the "nosologic tower of Babel" leitmotif which runs through any consideration of dementia and Lewy bodies. Harrington et al. found that APOE-4 allele frequency was increased in SDLT (0.365) and AD (0.328) compared with controls (0.147), PD (0.098), or Huntington's chorea (0.171) (Harrington, 1994). Nalbantoglu et al. similarly reported an analysis of APOE-4 genotype distribution in AD, AD/PD, and PD indicating that the APOE-4 carrier frequency was similar in the AD (0.60) and AD/PD groups (0.42), whereas the APOE-4 allele was absent in the PD group (Nalbantoglu et al., 1994). Still further confirmation of this finding came from Pickering-Brown et al. who reported a significant increase in the frequency of APOE-4 in what they termed "cortical Lewy body disease" as well as in AD compared with controls (Pickering-Brown, 1994). Benjamin et al. found increased APOE-4 frequency in both AD and SDLT but not in PD (Benjamin, 1994). Finally, St. Clair et al. (1994) also reported that what they refer to as "Lewy body disease" and AD groups had remarkably similar APOE-4 allelic distributions.

In a subsequent analysis of APOE-4 status and LBV we discovered that LBV brains from patients with one or two APOE-4 alleles had more advanced neurofibrillary pathology (i.e., a higher AD pathology stage), than LBV brains lacking this allele (Hansen, 1994).

Neurotransmitters in LBV

Unlike the results of the preceding neuropathologic analysis and APOE-4 genotyping, neurotransmitter deficiencies in LBV seem to link this entity more closely to PD than AD. We measured concentrations of the monoamines, their precursors, and their metabolites and the activity of choline acetyltransferase in basal ganglia and cortical regions of LBV, and normal controls (Langlais, 1993). Dopamine and homovanillic acid (HVA) were severely depleted in basal ganglia of the LBV cases, but not significantly altered in pure AD. Tyrosine hydroxylase levels in putamen were also significantly reduced in LBV but not AD. These reductions in basal ganglia dopamine and HVA suggest that LBV cases have a level of dopamine depletion similar to PD. Additionally, choline acetyltransferase activity in caudate and norenephrine concentrations in putamen are significantly reduced in LBV, which may contribute to the absence of resting tremor and the milder presentation of parkinsonian features in LBV compared to PD.

Antemortem diagnosis of LBV

Without straying too far afield from the neuropathological emphasis of the present study, it is important to emphasize that LBV is a clinical as well as a pathologic entity. As such, it can often be successfully diagnosed antemortem. Details concerning the clinical and psychometric manifestations of LBV are available in previous publications coming from the San Diego ADRC (Galasko, 1995). In brief, the best individual discriminators between AD and LBV are bradykinesia, rigidity, mask-like fascies, and parkinsonian gait, which are detected in about 30–50% of the LBV patients. Delusions are relatively more common in LBV (43%) than in AD (29%), although this does not attain statistical significance. Hallucinations, however, are significantly more frequent in LBV (32%) than in AD (11%), and they are always visual in nature. Neuropsychological test scores also help differentiate AD from LBV subjects. LBV patients show significantly greater impairment on tests of visuospatial and constructional abilities (Copy of Cube, Copy of Clock, Block Design Test), psychomotor speed (trails A), and verbal fluency.

Conclusions

Figure 3 illustrates the relationship LBV bears to AD on the one hand, and pure Lewy body disease, either PD or DLBD, on the other. All neuropathologists should agree that a brain with Lewy body disease which also shows many neocortical neuritic plaques and neurofibrillary tangles (neocortical "plaque and tangle" AD, equivalent to pathology stages V and VI), warrants some nosologic acknowledgement of concomitant Alzheimer disease. The plethora of synonyms for LBV, however, testifies to the nomenclature controversy which erupts when brains with Lewy bodies show only modest AD pathology (neocortical "plaque only or plaque predominant" AD, equivalent to pathology stages I-IV), represented by the gray shaded crescent in Fig. 3. Since both NIA and CERAD criteria for AD or probable AD are met by cases in this gray area, our preferred designation is Lewy body variant of Alzheimer disease. Other investigators, however, do not include AD in their nomenclature when referring to brains in this gray zone, preferring designations such as SDLT, DLBD, or cortical Lewy body disease. These alternative terminologies do not lend themselves to the distinction we make between LBV and pure DLBD. The increased APOE-4 allelic frequency in LBV supports our contention that the neocortical plaques and entorhinal tangles in LBV do represent AD and not coincidental age-related pathologic change.

Subcortical and neocortical Lewy body pathology also contribute substantively to the clinical and pathologic entity we call LBV. Lewy body pathology presumably causes the mild parkinsonism, visual hallucinations, and characteristic neuropsychometric abnormalities which distinguish LBV from AD during life. More speculatively, it may be that LBV, representing about 25% of all clinically diagnosed AD, accounts for the minority of responders to treatment of AD with tetrahydroaminoacridine (tacrine). This hypothesis seems feasible in light of the greater neocortical choline acetyltransferase deficits in LBV compared to AD, and the cholinesterase inhibition induced by tacrine treatment.

Acknowledgements

This research was supported financially by a grant from the NIH #2P50 AG05131-11 Alzheimer's Disease Research Center. The author wishes to publically express his gratitude to P. Zuniga for her efficiency and expertise in preparing this manuscript.

References

Alford MF, Masliah E, Hansen LA, Terry RD (1994) A simple dot- immunobinding assay for quantification of synaptophysin-like immunoreactivity in human brain. J Histochem Cytochem 42: 283–287

Benjamin R, Leake A, Edwardson JA, McKeith IG, Ince PG, Perry RH, Morris CM (1994) Apolipoprotein E genes in Lewy body and Parkinson's disease. Lancet 343: 1565

Bergeron C, Pollanen M (1989) Lewy bodies in Alzheimer disease; one or two diseases? Alzheimer Dis Assoc Disord 3: 197–204

Bierer LM, Perl DP, Haroutunian V, Mohs RC, Davis KL (1990) Neurofibrillary tangles, Alzheimer's disease, and Lewy bodies. Lancet 335 8682: 163

Braak H, Braak E (1991) Neuropathological stageing of Alzheimer-related changes. Acta Neuropathol 82: 239–259

Dickson DW, Davies P, Mayeux R, Crystal H, Horoupian DS, Thompson A, Goldman JE (1987) Diffuse Lewy body disease: neuropathological and biochemical studies of six patients. Acta Neuropathol 75: 8–15

Ditter SM, Mirra SS (1987) Neuropathologic and clinical features of Parkinson's disease in Alzheimer's disease patients. Neurology 37: 754–60

Galasko D, Saitoh T, Xia Y, Thal LJ, Katzman R, Hill LR, Hansen L (1994) The apolipoprotein E allele e4 is over-represented in patients with the Lewy body variant of Alzheimer's disease. Neurology 44: 1950–1953

Galasko K, Katzman R, Salmon DP, Thal LJ, Hansen L (1996) Clinical and neuropathological findings in Lewy body dementias. Brain Cogn 31: 166–175

Hansen LA, Masliah E, Terry RD, Mirra SS (1989) A neuropathological subset of Alzheimer's disease with concomitant Lewy body disease and spongiform change. Acta Neuropathol 78: 194–201

Hansen L, Salmon D, Galasko D, Masliah E, Katzman R, DeTeresa R, Thal L, Pay M, Hofstetter R, Klauber M, Rice V, Butters N, Alford M (1990) The Lewy body variant of Alzheimer's disease: a clinical and pathological entity. Neurology 40: 1–8

Hansen LA, Masliah E, Quijada-Fawcett S, Rexin D (1991) Entorhinal neurofibrillary tangles in Alzheimer's disease with Lewy bodies. Neurosci Lett 129: 269–272

Hansen LA, Masliah E, Galasko D, Terry RD (1993) Plaque only Alzheimer disease is usually the Lewy body variant, and vice versa. J Neuropathol Exp Neurol 52: 648–654

Hansen LA, Galasko D, Samuel W, Xia Y, Chen X, Saitoh T (1994) Apolipoprotein-E e-4 is associated with increased neurofibrillary pathology in the Lewy body variant of Alzheimer disease. Neurosci Lett 182: 63–65

Harrington CR, Louwagie J, Rossau R, Vanmechelen E, Perry RH, Perry EK, Xuereb JH, Roth M, Wischik CM (1994) Influence of apolipoprotein E genotype on senile dementia of the Alzheimer and Lewy body types. Am J Pathol 145: 1472–1484

Ince P, Irving D, MacArthur F, Perry RH (1991) Quantitative neuropathological study of Alzheimer-type pathology in the hippocampus: comparison of senile dementia of Lewy body type, Parkinson's disease and non-demented elderly control patients. J Neurol Sci 106: 142–52

Joachim CL, Morris JH, Selkoe DJ (1988) Clinically diagnosed Alzheimer disease: autopsy results in 150 cases. Ann Neurol 24: 50–6

Khachaturian ZS (1985) Diagnosis of Alzheimer's disease. Arch Neurol 42: 1097–105

Kosaka K (1990) Diffuse Lewy body disease in Japan. J Neurol 237: 197–204

Langlais PJ, Thal L, Hansen LA, Galasko D, Alford M, Masliah E (1993) Neurotransmitters in basal ganglia and cortex of Alzheimer's disease with and without Lewy bodies. Neurology 43: 1927–1934

Menzel H, Kladetzky RG, Assman G (1983) Apolipoprotein E polymorphism and coronary artery disease. Arteriosclerosis 3: 310–315

Mirra SS, Hart MN, Terry RD (1993) Making the diagnosis of Alzheimer's disease. Arch Pathol Lab Med 117: 132–144

Nalbantoglu J, Gilfix BM, Bertrand P, Robitaille Y, Gauthier S, Rosenblatt DS, Poitier J (1994) Predictive value of apolipoprotein E genotyping in Alzheimer's disease: results of an autopsy series and an analysis of several combined studies. Ann Neurol 36: 889–895

Perry RH, Irving D, Blessed G, Fairbairn A, Perry EK (1990) Senile dementia of the Lewy body type. A clinical and neuropathologically distinct form of Lewy body dementia in the elderly. J Neurol Sci 95: 119–139

Pickering-Brown SM, Mann DMA, Bourke JP, Roberts DA, Bilderson D, Burns A, Byrne J, Owen F (1994) Apolipoprotein E4 and Alzheimer's disease pathology in Lewy body disease and in other beta amyloid forming diseases. Lancet 343: 1115

St. Clair D, Norrman J, Perry R, Yates C, Wilcock G, Brooks A (1994) Apolipoprotein E e4 allele frequency in patients with Lewy body dementia, Alzheimer's disease and age-matched controls. Neurosci Lett 176: 45–46

Strittmatter WJ, Saunders AM, Schmechel D, Pericak-Vance M, Enghild J, Salvesen GS, Roses AD (1993) Apolipoprotein E: high-avidity binding to β-amyloid and increased frequency of type 4 allele in late-onset familial Alzheimer disease. Proc Natl Acad Sci USA 90: 1977–1981

Terry RD, Hansen LA, DeTeresa R, Davies P, Tobias H, Katzman R (1987) Senile dementia of the Alzheimer type without neocortical neurofibrillary tangles. J Neuropathol Exp Neurol 46: 262–268

Author's address: L. A. Hansen, M.D., Departments of Pathology and of Neurosciences, School of Medicine, University of California, San Diego, 9500 Gilman Drive, La Jolla, Ca 92093-0624, U.S.A.

Lewy body dementia — clinical, pathological and neurochemical interconnections*

R. Perry[1], I. McKeith[2], and E. Perry[3]

[1]Neuropathology Department, and [2]Department of Old Age Psychiatry, University of Newcastle, and [3]Medical Research Council, Neurochemical Pathology Unit, Newcastle General Hospital, Newcastle upon Tyne, United Kingdom

Summary. Senile dementia of Lewy body type or Lewy body dementia (SDLT or LBD) is defined as a Lewy body associated disease presenting in the elderly primarily with dementia with variable extrapyramidal disorder. Characteristic clinical symptoms include fluctuating cognitive impairment, psychotic features such as hallucinations and a particular sensitivity to neuroleptic medication. Although apolipoprotein e4 allele is increased 2–3 fold in SDLT (as in Alzheimer's disease) and β-amyloidosis occurs in most cases, the most robust neurobiological correlate of the dementia so far identified appears to be extensive cholinergic deficits in the neocortex. This is consistent with previously reported correlations between cortical cholinergic activity and dementia in Parkinson's disease (PD) and Alzheimer's disease. There is also a significant interaction between the density of limbic cortical Lewy bodies and dementia in both SDLT and PD, although the cortical neuronal population affected remains to be identified. Cortical Lewy body density is positively correlated with the age of disease onset in PD and SDLT. This may account for the increased incidence of psychiatric syndromes, as opposed to extrapyramidal disorder in Lewy body disease with advancing age as may age-related loss of cholinergic activity in cortical areas such as the hippocampus.

Clinical syndromes associated with cortical Lewy body disease

Patients with Parkinson's disease (PD) frequently have mild cognitive impairment, a 2 to 4 times increased relative risk of developing dementia and are also susceptible to episodic confusional states. These symptoms pre-date the L-dopa era (Woodard, 1962). Extrapyramidal features have also been

---

* Since this paper was prepared, the Consortium for Dementia with Lewy bodies (CDLB) met in 1995 and established consensus clinical diagnostic criteria and pathological guidelines for the identification of Dementia with Lewy bodies [published in McKeith, et al (1996) Neurology 47: 1113–1124]

recognised in up to 30% of clinically diagnosed Alzheimer's disease (AD) patients and are predictive of reduced survival. Despite this overlap, PD and AD continue to be considered as nosologically separate clinical entities.

The emerging model of a pathological spectrum (Perry et al., 1990a) challenges this clinical dichotomy. Attempts to clarify the relation between AD and PD have involved researchers retrospectively reviewing the clinical features, both psychiatric and neurological, of small samples of patients, identified as occupying an intermediate point on such a pathological spectrum. These attempts characterised by Burkhardt et al. (1988), Crystal et al. (1990) and more recently culminating in the Nottingham (Byrne et al., 1991) and Newcastle (McKeith et al., 1992a) operationalised criteria, may be likened to the story of the 6 blind men and an elephant. Only by pulling together their individual perceptions can the true nature of the beast be characterised. Much has been made of the biases produced by different sampling frames — some patients coming from predominantly neurology referred populations and others sampled from psychiatric clinics. An equally valid and perhaps more important consideration is that the absolute number of patients considered is very limited. Byrne (1995) refers to about 300 reported cases of cortical Lewy body disease — in less than half of these is there sufficient clinical data available to be able to make even general statements about the pattern of clinical features. Table 1 shows an analysis of all such cases published to the end of 1992. This review suggests that the clinical heterogeneity associated with cortical Lewy body pathology is truly diverse, approximately 25% of patients initially presenting with motor features of Parkinson's disease, 50% presenting with cognitive impairment only and a substantial minority presenting with other symptoms including delusional states (variously diagnosed as late paraphrenia/delusional disorder or major depression with psychosis) or physical symptoms including dizziness and falls.

Clinical features of senile dementia of Lewy body type

The clinical syndrome reflected in the original Newcastle description (Perry et al., 1989, 1990a) and clinical diagnostic criteria for SDLT (McKeith et al.,

Table 1. Literature review of 139 published cases identified 1961–1992 in which clinical details are available for patients with diffuse (cortical) Lewy body disease (n = 139)

	Symptoms at onset of illness	Symptoms at late stage of illness
Cognitive impairment only	72 (52%)	28 (20%)
Parkinsonism only	24 (17%)	5 (4%)
Parkinsonism and cognitive impairment	11 (8%)	106 (76%)
Others*	21 (15%)	
Not recorded	11 (8%)	

*10 (48%) = psychosis, 6 (29%) = dizziness/falls, 3 (14%) = depression, 1 = headache, 1 = personality deterioration

Table 2. Operational criteria for senile dementia of Lewy body type (McKeith et al., 1992b)

A, B, C, D and E must be fulfilled

A — Fluctuating cognitive impairment affecting both memory and higher cortical functioning (such as language, visuospatial abilities, praxis or reasoning skills). The fluctuation is marked with the occurrence of both episodic confusion and lucid intervals, as in delirium, and is evident either on repeated tests of cognitive function or by a variable performance in daily living skills.
B — At least one of the following:
    1. Visual and/or auditory hallucinations which are usually accompanied by secondary paranoid delusions
    2. Mild spontaneous extrapyramidal features or neuroleptic sensitivity syndrome, i.e. exaggerated adverse responses to standard doses of neuroleptic medication
    3. Repeated unexplained falls and/or transient clouding or loss of consciousness
C — Despite the fluctuating pattern, the clinical features persist over a long period of time (weeks or months), unlike delirium which rarely persists long. The illness progresses, often rapidly, to an end stage of severe dementia
D — Exclusion of any underlying physical illness adequate to account for the fluctuating cognitive state, by appropriate examination and investigation
E — Exclusion of past history of confirmed stroke and/or evidence of cerebral ischaemic damage on structural brain imaging

---

1992a,b; Table 2) are derived from cases taken from a sampling frame favouring psychiatric presentations progressing to dementia (but with no requirement to meet any specified clinical pattern) and with a slight negative bias towards patients with established Parkinson's disease (these patients being referred to separate neurological services). Tables 3 and 4 illustrate the symptom profile of the Newcastle SDLT patients with comparative data for an Alzheimer's disease group. Twenty per cent (4/20) of Lewy body patients had previously been diagnosed as having motor Parkinson's disease — a figure similar to that obtained by literature review. A further 25% (5/20) had extrapyramidal features at the time of presentation to specialist psychiatric services which were attributable to neuroleptic medication. The characteristics of some of the Lewy features of SDLT are briefly described below.

Fluctuating cognitive impairment

This refers to fluctuations in cognitive performance not attributable to co-existing systemic disease or drug toxicity. In some patients it is evident as discrete "confusional episodes" accompanied by disorientation, clouding of consciousness, impaired concentration, distractibility and often associated with perceptual disorders including misidentification syndromes and hallucinations. Fluctuation may also be apparent in periods of hypoactivity and somnolence. Cognitive fluctuation has proved difficult to quantify. Individual patient's performances on standardised mental tests may vary dramatically within relatively short periods of time, in contrast to Alzheimer's disease patients where scores remain relatively stable. Sahgal et al. (1992) have shown

Table 3. Newcastle SDLT cases 1990–1992

	SDLT (n = 20)		AD (n = 21)	
	At presentation	Occurring over	At presentation	Occurring over
Extrapyramidal features	9 (45)	16 (75)	1 (5)	4 (19)
Previous diagnosis of PD	4 (20)	4 (20)	0	0
EPF secondary to neuroleptics	1 (5)	7 (45)	1 (5)	4 (19)

Table 4. Newcastle SDLT cases 1990–1992

	SDLT (n = 20)		AD (n = 21)	
	At presentation	Occurring ever	At presentation	Occurring ever
Fluctuating cognitive impairment	17 (85)*	18 (90)*	1 (5)	1 (5)
Visual hallucinations	11 (55)*	16 (80)*	4 (19)	4 (19)
Auditory hallucinations	6 (30)*	9 (45)*	0	0
Delusions	13 (65)*	16 (80)*	4 (19)	4 (19)
Repeated unexplained falls	7 (35)	10 (50)*	3 (14)	5 (24)
Transient disturbances of consciousness	5 (25)	5 (25)	1 (5)	5 (24)

*p < 0.05 compared with AD at similar stage

qualitative differences between SDLT and AD patients on the attentional set shifting paradigm of CANTAB, similar to the deficit seen in PD, perhaps reflecting subcortico-frontal loop dysfunction.

Psychotic symptoms

Visual hallucinations are seen in the majority of SDLT patients. They often occur as part of the presenting symptomatology and are persistent throughout the illness, even to the late stages. Typically they are well formed, detailed, colourful, moving images, auditory hallucinations occur less frequently and are usually transient. They seldom occur alone. Patients occasionally report tactile and olfactory hallucinations. Hallucinatory experiences may occur particularly on waking from sleep and are often difficult to distinguish from nightmares. Delusions almost always relate to the content of hallucinations.

Falls and disturbances of consciousness

Fifty per cent of SDLT patients have repeated unexplained falls and it is likely that a substantial number present to geriatric medicine services in this way.

Three possible mechanisms have been postulated — postural hypotension, gait and balance problems or a primary cerebral aetiology. In this latter context, it is of interest that 50% of SDLT patients have transient slow wave (delta range) activity in the temporal lobes.

Neuroleptic sensitivity

Eighty-one per cent of SDLT patients prescribed neuroleptics showed neuroleptic sensitivity (McKeith et al., 1992b). In 50% of cases this was mild (the development or worsening of parkinsonian features following medication in the accepted dose range for elderly patients) and in 50% it was severe (acute and severe physical deterioration for which no other adequate cause was apparent and which appeared related in time to neuroleptic prescription). Survival times from presentation to psychiatric services until death for patients with severe neuroleptic sensitivity were reduced to 9.6 months (95% CI, 2.6–16.5) compared with 26.7 months (95% CI, 3.3–50.0) for those not neuroleptic sensitive and 25.0 months (95% CI, 0.6–49.4) for those not exposed to neuroleptics, unpaired t-test, $t = -1.91$, $p = 0.035$ (one tailed). A survival analysis (McKeith et al., 1992b) showed an increased mortality risk (expressed as a hazard ratio of 2.70) during the first 12 months after presentation in patients with severe neuroleptic sensitivity. The neurochemical basis of this is discussed below.

Epidemiology

Aside from the issue of the correlation between the clinical syndrome which we have described as SDLT and its relationship to a point or points on the pathological spectrum represented in Fig. 1, there are at least 2 major clinical issues which need to be addressed. First there is the question of inter-rater reliability of clinical diagnosis and secondly the issue of prevalence of the clinical syndrome in representative clinical samples.

To address the first question a range of clinical diagnostic criteria for dementia subtypes were applied to the case notes of autopsy confirmed SDLT (n = 20), AD (n = 21) and multi-infarct dementia (MID, n = 9) patients who had received comprehensive psychogeriatric assessment (McKeith et al., 1994a). The inter-rater reliability for the clinical diagnosis of SDLT varied between 4 raters from k = 0.50–0.88 (mean = 0.64) indicating a substantial level of agreement similar to that obtained for existing clinical methods of diagnosing Alzheimer's disease. Diagnostic specificity for SDLT was uniformly high (0.9–0.97) indicating a low false-positive diagnosis rate and the mean sensitivity of detection was 0.74, substantially greater (0.9) for an experienced, compared with a relatively inexperienced (0.55) rater. This suggests that the antemortem identification of SDLT patients can be achieved with a high degree of specificity, although there remains a substantial minority of patients who are difficult to detect because of presentations with either "typical" Alzheimer-type symptoms or with paranoid or delusional symptoms in the absence of substantial cognitive impairment.

Applying other sets of clinical criteria to SDLT patients, the most probable erroneous diagnosis was one of Alzheimer's disease (15% by NINCDS "probable AD", 35% by DSMIIIR dementia of Alzheimer type, and 50% by NINCDS "possible AD").

Two recent studies have measured the prevalence of the clinical syndrome of SDLT. Ballard et al. (1993) found that 14/58 (24%) of psychogeriatric day hospital attenders with a diagnosis of dementia fulfilled Newcastle criteria for SDLT, 94% having frequent falls, 25% visual hallucinations, 69% extrapyramidal symptoms and 50% a family history of Parkinson's disease. Shergill et al. (1994) applied Newcastle criteria for SDLT and Nottingham criteria for Lewy body dementia to 114 patients referred to an old age psychiatry service with a primary diagnosis of dementia. Twenty-six per cent fulfilled Newcastle criteria and a total of 23.7% the Nottingham criteria. The tendency to clinically misdiagnose patients with Lewy body dementia as having Alzheimer's disease was confirmed. This is strong supportive evidence that the syndrome which has been described as characteristic of SDLT is relatively common in clinical practice and accords with the prevalence figures quoted for Lewy body dementia in autopsy series.

Neuropathological features

Dementia associated with Lewy bodies (SDLT and PD developing dementia) may be multifactorial — related neuropathologically in different subgroups to the presence of β-amyloid plaques, neurofibrillary tangles or the presence of cortical Lewy bodies. An alternative hypothesis is that there are common pathogenetic mechanisms underlying the dementia. In searching for a unitary neurobiological basis of dementia in SDLT, features common to all cases thus need to be considered.

β-Amyloidosis occurs in most *but not all* cases of LBD (Perry et al., 1990a). In a recent series β-amyloidosis was increased in 7 out of 21 (32%) demented compared with 3 out of 14 (21%) non-demented cases of Parkinson's disease (PD). This increase in β-amyloid density was accompanied by scattered neocortical tangle formation (at densities between 10 and 20 × lower than AD) in 5 cases. The pathology in approximately one-third of demented PD cases therefore resembles that seen in the majority of SDLT cases (i.e. raised neocortical amyloid density but few or absent neocortical tangles). Only one demented Parkinson's case had coexisting Alzheimer's disease as defined by the presence of high density of neocortical plaques and tangles. Analysis of this series does not suggest that dementia in PD is primarily related to coexisting AD.

The apolipoprotein e4 allele is present in just over half of SDLT cases and not significantly different from the normal in PD, in which group non-demented and demented cases are similar (Table 5) (Harrington et al., 1994).

Table 5. Apolipoprotein E allele distributions

Category[a]	Case numbers	Apolipoprotein E alleles					
		2′2	2′3	2′4	3′3	3′4	4′4
Controls	40	0	3 (7.5%)	3 (7.5%)	25 (62.3%)	9 (22.5%)	0
Parkinson's disease[b]	33	0	5 (15%)	1 (3%)	15 (45%)	12 (36%)	0
Senile dementia of Lewy body type	29	0	2 (6%)	3 (10.3%)	8 (27.5%)	13 (45%)	3 (10.3%)
Alzheimer's disease	51	0	0	3 (6%)	16 (31.4%)	20 (39%)	12 (23.5%)

[a] Cases were allocated on the basis of clinical and pathological criteria previously described (Perry et al., 1990a). [b] Amongst 9 non-demented Parkinson cases there were 4 with 393 and 5 with 394; in 18 demented cases there were 2 with 293, 10 with 393 and 5 with 394

In contrast to Alzheimer-type pathology and genotype, cortical Lewy bodies occur (by definition) in all SDLT cases, and to a lesser extent in all cases of idiopathic PD (Table 6). In LBD, as in PD, cortical Lewy bodies are concentrated in limbic areas (e.g. cingulate and parahippocampal or entorhinal cortex). There is a relationship between Lewy body density in these areas and dementia. Numbers are significantly higher in the demented compared to non-demented PD subgroups in anterior cingulate and parahippocampal gyrus (Table 6) and also higher in SDLT compared with PD in these limbic areas (Perry et al., 1990a).

Although cortical Lewy bodies are relatively most dense in limbic cortical areas, numbers are in absolute terms nevertheless very low. For example a density of 50 Lewy bodies per cm^2 in limbic cortex compares with plaque or tangle densities of over 20 per mm^2 in Alzheimer's disease. Unless there is a minor sub-population of distinct and functionally relevant cortical neurons which are affected by Lewy bodies, the presence of these pathological features in the cortex may be no more than diagnostic and indicative of more widespread pathology, particularly affecting key subcortical areas such as nucleus basalis of Meynert.

A key morphological correlate of dementia may be de-afferentation of the cortex related to degeneration of cortically projecting subcortical nuclei. This would be consistent with recent imaging analyses which indicate reduced cortical glucose metabolism in LBD and PD, especially those with dementia (see Brooks, this issue). An obvious candidate for degenerating cortical input is cholinergic axons from the basal forebrain system. Morphologically, loss of cholinergic axons is apparent in cortical sections stained for acetylcholinesterase histochemical activity (unpublished observations). The cholinergic input to archicortical areas such as amygdala and hippocampus is particularly dense, especially in the CA2/3 region of the hippocampus. It is this area in which distinct, abnormal, ubiquitinated neurites not evident in Alzheimer's

Table 6. Prevalence of cortical Lewy bodies in Parkinson's disease

| | Cases showing Lewy bodies in specific cortical regions | | | | | | | | |
| | Neocortex | | | | | Limbic cortex | | | |
	Frontal	Temporal	Parietal	Occipital	Mean	Anterior Cingulate	Posterior Cingulate	Hippocampal Gyrus	Mean
Non-demented (10)	40%	40%	0%	0%	50%	70%	40%	80%	100%
Demented (14)	57%	64%	21%	0%	79%	93%*	73%	93%*	100%

* Significant difference non-demented and demented ($p < 0.02$)

disease have been identified in LBD (Dickson et al., 1991). This neurohistological feature may be directly linked to cortical cholinergic degeneration and the identification of such ubiquitinated neurites in the CA1/CA2 region of non-demented Parkinson's disease cases (Perry RH, unpublished observation) may be associated with cholinergic dysfunction in this group of cases. Cholinergic deafferentation of the cortex in LBD may thus involve a degenerative process distinct from that which occurs in Alzheimer's disease in which there is evidence for dysfunctional as opposed to degenerative cortical cholinergic abnormalities (Perry et al., 1993).

Neurochemical/pharmacological aspects

One of the cardinal clinical features of SDLT is the fluctuating nature of the cognitive impairment (Section II). Although the disease is progressive there are, at least in its early stages, periods of remission when cognition returns to near normal. This scenario is less compatible with irreversible cellular neurodegeneration and more suggestive of reversible functional changes. Identification of the nature of such functional changes requires high resolution, chemically tagged scanning devices applied sequentially to patients throughout the course of their disease. Autopsy biochemical data is clearly limited in time by its single point/cross sectional, end stage observation. Nevertheless there are intriguing differences in the neurochemical pathology of LBD and AD, identified in brain bank material, which suggest that involvement of the cortical cholinergic system may provide important clues as to the clinical nature of SDLT.

Although neurochemical analysis of brain tissue in SDLT has been limited to date, in the few published reports on neocortical cholinergic activity (Dickson et al., 1987; Perry et al., 1994) a more extensive loss of neocortical choline acetyltransferase (ChAT) occurs than is seen in AD. This is particularly so in parietal neocortex (Fig. 1). Neocortical temporal and parietal activities are similar in LBD and in demented cases of Parkinson's disease. In contrast hippocampal cholinergic activities are generally lower in AD than SDLT (Fig. 1).

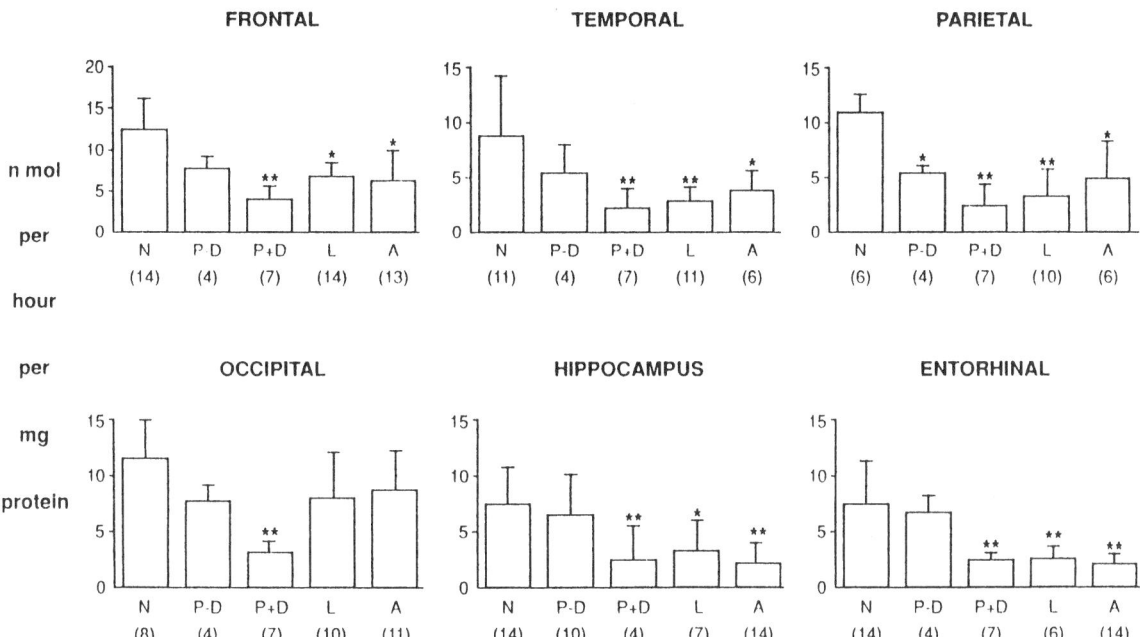

Fig. 1. Choline acetyltransferase activities in different cortical regions in normal (N), Parkinson's disease without (P − D) and with (P + D) dementia, Lewy body dementia (L) and Alzheimer's disease (AD). Columns represent mean values, bars standard deviations and numbers are case numbers. Significant (t-test) differences from the normal are marked *(p, 0.02) and **(p < 0.001)

These neurochemical data are anatomically consistent with the respective clinical features of SDLT and AD. Memory impairment is a consistent, early feature of AD whereas psychotic features such as hallucinations are a consistent, early feature of many SDLT cases. SDLT patients experiencing hallucinations have lower neocortical ChAT than those not reporting this experience (Perry et al., 1990b). This neurochemical finding is consistent with the widely reported effects of anticholinergic agents (such as scopolamine or atropine) to induce hallucinations (Perry et al., 1995). Parkinson patients exhibit increased sensitivity to anticholinergics — in terms of induced memory loss or delirium — compared with age-matched normal individuals. In addition L-dopa can induce hallucinations in PD. This raises the question of whether hallucinations depend on a hypermonoaminergic — hypocholinergic imbalance. This has been identified in SDLT, in relation to the 5-HT system (Perry et al., 1990b).

There are also intriguing interactions between the cholinergic and dopaminergic systems which may account for the lack of overt dementia or other cognitive impairment in PD cases despite quite extensive cholinergic deficits in some areas (Fig. 1). Thus, scopolamine-induced deficits in spatial working memory in rodents are reversed by dopaminergic D_1 receptor blockade by SCH 23390 (Levin et al., 1990). The D_1 receptor appears to inhibit acetylcholine release in the hippocampus through its action in the septum (Durkin et al., 1986). Thus degeneration of ventral tegmental dopaminergic neurons in PD may "offset" the detrimental effects of basal forebrain cholinergic degeneration. A similar argument applies, in relation to basal ganglia circuitry, to the protective effects of therapeutic cholinergic blockade (via antimuscarinics) on the functional consequences of substantia nigra dopaminergic neuron loss in terms of extrapyramidal movement disorder. We have recently identified a case of corticobasal degeneration with <5% surviving substantia nigra dopaminergic neurons and <1% of the mean normal level of putamen dopamine. In this case, typical clinical features of parkinsonism were not apparent, a paradox that may be attributable to the extensive cholinergic deficit in the putamen such as is not seen in classical PD (Marshall et al., in preparation).

Common features of the cortical cholinergic involvement in SDLT and PD include not only the extensive neocortical deficits but also supersensitivity (in terms of numbers of binding sites) of the muscarinic receptor population (Perry et al., 1990c). Together with the general absence of neocortical tangles, the neurochemical data indicate that patients with SDLT may be more responsive to cholinergic therapy than classical cases of AD (Perry et al., 1989). Apart from preliminary reports on the identification of a small number of tacrine responders with Lewy bodies in two tacrine trials with autopsy data to date (Levy et al., 1994; Wilcock and Scott, 1994) this proposal remains to be verified clinically. Tacrine, in addition to its potential liver toxicity, is restricted in its therapeutic utility due to interactions with all CNS cholinergic systems. In SDLT and particularly cognitively impaired PD patients enhancing striatal cholinergic transmission could exacerbate extrapyramidal movement disorder. There is a great need for regionally targeted cholinergic therapy to treat not only the memory disorder of AD but also some of the non-cognitive or psychotic features associated with e.g. SDLT.

Typical neuroleptics which involve dopaminergic receptor D_2 blockade are, although effective in reducing psychosis e.g. hallucinations, clearly inappropriate on account of the severe parkinsonism induced in many SDLT cases (Section I). Neurochemical analysis of the striatum in LBD patients intolerant of neuroleptic medication indicates a failure of the D_2 upregulation mechanisms in response to receptor blockade (Piggott et al., 1994). Comparing the putamen in intolerant and tolerant cases, levels of dopamine and its metabolites, D_1 receptor binding and cholinergic activity were all similar, but D_2 receptor numbers were increased in the tolerant (as in PD cases) but not the intolerant individuals (Piggott et al., in preparation). Atypical neuroleptics such as clozapine with a lower affinity for D_2 compared with other monoaminergic receptors such as dopaminergic D_3, 5-HT_2 and even muscarinic receptors may be of value in SDLT therapy. Clozapine has recently been reported to have a high affinity for and act as an agonist at the muscarinic m_4 subtype (Zorn et al., 1994). This receptor predominates in the striatum and is also relatively high in the cortex. Since muscarinic receptor molecular biology has recently advanced with the definition of over 5 different subtypes, there are new opportunities for examining the status of these in LBD and PD both in cortex and striatum. In addition novel muscarinic agents targeted to specific subtypes may extend specific applications of cholinergic therapy. In cloned muscarinic receptors (Buckley et al., 1989), subtypes m_1, m_2, m_4 have a higher affinity compared with m_2 for atropine (one of the tropane alkaloids which can induce psychosis reminiscent of LBD). Therapy with cholinesterase inhibitors such as tacrine, which elevate acetylcholine levels, will presumably involve stimulation of all muscarinic subtypes, including m_2 and it may be that selective muscarinic receptor stimulation is more appropriate.

Ageing may be a critical factor in clinical presentation

The extensive involvement of the basal forebrain cholinergic system projecting to the cortex may be one of the most robust correlates of dementia in SDLT. This condition, presenting in the elderly initially with dementia and subsequently variable or minimal extrapyramidal disorder shares this neurochemical feature with classical Parkinson's disease cases developing dementia. It seems likely that the two categories belong to the same disease spectrum. The increased incidence of dementia as opposed to the classical extrapyramidal disorder of PD with increasing age (beyond 70–80 years) may reflect the occurrence of independent age-associated changes such as increased density of cortical Lewy bodies, the appearance of β-amyloid or degeneration of the cortical cholinergic system, which predispose to cognitive impairment.

Lewy body densities in limbic cortex are significantly and positively correlated with age at presentation in a recent series of neuropathologically assessed cases of SDLT and PD (Fig. 2). However, no biochemical marker of a select population of cortical neurons, related to the presence of Lewy bodies in the cortex, has yet been determined which identifies the distinct neurochemical pathology of LBD. Moderate decrements in somatostatin and CRF neuropeptide immunoreactivities, similar to those occurring in AD, have

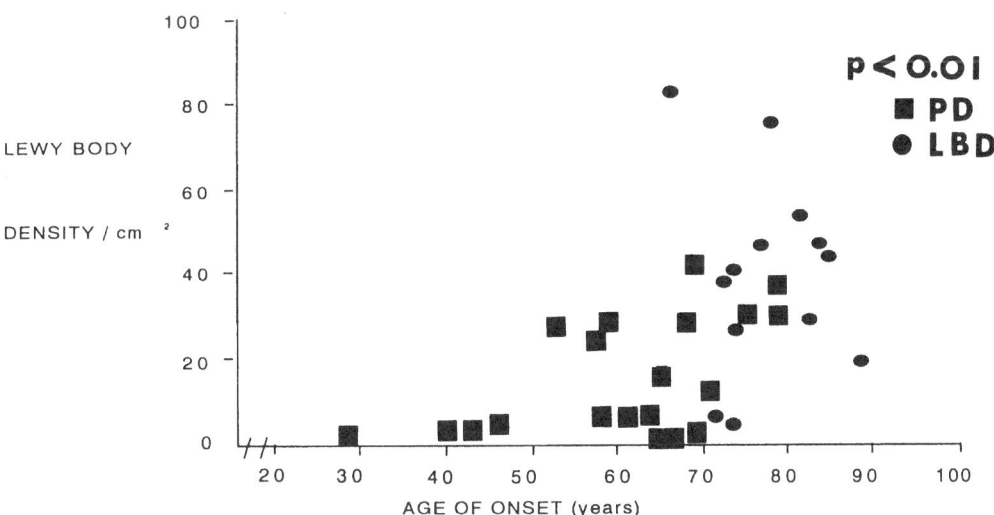

Fig. 2. Density of Lewy bodies in anterior cingulate cortex in patients with Parkinson's disease (■) and senile dementia of Lewy body type (●) as a function of age of onset of clinical symptoms. Correlation coefficient significant ($p < 0.01$)

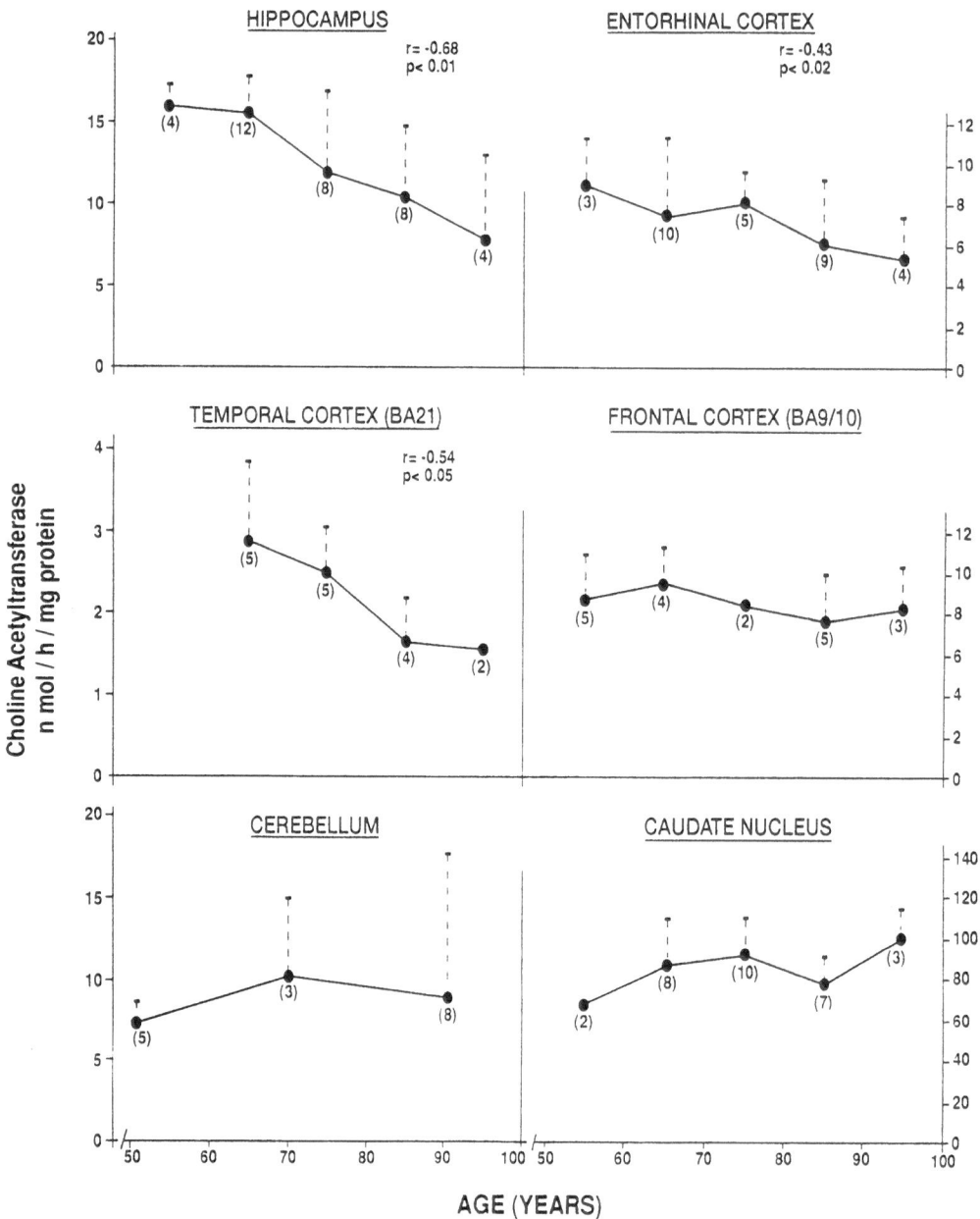

Fig. 3. Choline acetyltransferase activities in different brain areas in normal individuals aged 50–100 years. Points represent mean values and bars standard deviations, case numbers in parentheses. Significant reductions across this age range occurred in hippocampus, entorhinal and temporal neocortex but not in the other areas [with permission from Perry (1997)]

been reported in LBD and PD. Since Lewy bodies do not occur in classical AD, in which the extent of peptide losses is similar to that in LBD, it is unlikely that these peptides are specifically involved in the dementia of SDLT.

Age-related alterations in cholinergic activity are generally reported in most studies of elderly human and non-human species, albeit in subgroups of the elderly population. Regional analysis in normal human brain (Fig. 3) has indicated that the loss of choline acetyltransferase is marked in hippocampus and other temporal cortical areas and not evident in non-cortical areas including striatum and cerebellum (Perry et al., 1977; Court et al., 1993). Whether dopaminergic activities decline with increasing age is unclear. Recent [18F] dopa imaging data are inconsistent, some studies reporting no change in striatal uptake with increasing age (Sawle, 1995). Lewy body associated diseases such as PD and SDLT involve progressive loss of both striatal dopaminergic and cortical cholinergic activities generally occurring in the elderly population. If pathological changes occur against a background of selective age-related cortical cholinergic degeneration this would provide an explanation for the shift in clinical presentation from extrapyramidal to cognitive disorder.

Clinical presentation may thus depend critically on age of presentation. SDLT is however associated not only with increasing age but also with increased apo e4 allele frequency — a risk factor for accelerating the onset of dementia in AD. The complex relation between this genotype, ageing and the clinical or pathological phenotype of dementia associated with Lewy bodies remains to be established.

Acknowledgements

The secretarial assistance of M. Middlemist and A. Nicholson are gratefully acknowledged.

References

Ballard CG, Mohan RNC, Patel A, Bannister C (1993) Idiopathic clouding of consciousness — do the patients have cortical Lewy body disease? Int J Geriatr Psychiatry 8: 571–576

Buckley NJ, Bonner T, Buckley CM, Brann MR (1989) Antagonist binding properties of five cloned muscarinic receptors expressed in CHO-Ki cells. Mol Pharmacol 35: 469–479

Burkhardt CR, Filley CM, Kleinschmidt-DeMasters BK, de la Monte S, Norenberg MD, Schneck SA (1988) Diffuse Lewy body disease and progressive dementia. Neurology 38: 1520–1528

Byrne EJ (1995) Cortical Lewy body disease: an alternative view. In: Levy R, Howard R (eds) Developments in dementia and functional disorders in the elderly. Wrightson, London, pp 21–30

Byrne EJ, Lennox G, Lowe J, Godwin-Austen RB, Jefferson D, Lowe J, Mayer RJ, Landon M, Doherty FJ (1991) Diagnostic criteria for dementia associated with cortical Lewy bodies. Dementia 2: 283–284

Court JA, Perry EK, Johnson M, Piggott MA, Kerwin JM, Perry RH, Ince PG (1993) Regional patterns of cholinergic and glutamate activity in the developing and aging human brain. Dev Brain Res 74(1): 73–82

Crystal HA, Dickson DW, Lizardi JE, Davies P, Wolfson LI (1990) Antemortem diagnosis of diffuse Lewy body disease. Neurology 40: 1523–1528

Dickson DW, Davies P, Mayeux R, Crystal H, Haroupian DS, Thompson J, Goldman JE (1987) Diffuse Lewy body disease. Acta Neuropathol 75: 8–15

Dickson DW, Ruan D, Crystal H, Mark MH, Davies P, Kress Y, Yen S-H (1991) Hippocampal degeneration differentiates diffuse Lewy body disease (DLBD) from Alzheimer's disease. Neurol 41: 1402–1405

Durkin T, Galey D, Micheau J, Beslon H, Jaffard R (1986) The effects of intraseptal injection of haloperidol in vivo on hippocampal cholinergic function in the mouse. Brain Res 376: 420–424

Harrington CR, Louwagie J, Rossau R, Vanmechelen E, Perry RH, Perry EK, Xuereb JH, Roth M, Wischik CM (1994) The influence of apolipoprotein E genotype on senile dementia of the Alzheimer and Lewy body types. Am J Pathol 145: 1472–1484

Levin ED, McGurk SR, Rose JE, Butcher LL (1990) Cholinergic-dopaminergic interactions in cognitive performance. Behav Neural Biol 54: 271–299

Levy R, Eagger S, Griffiths M, Perry EK, Honavar M, Dean A, Lantos P (1994) Lewy bodies and response to tacrine in Alzheimer's disease. Lancet 343: 176

McKeith IG, Perry RH, Fairbairn AF, Jabeen S, Perry EK (1992a) Operational criteria for senile dementia of Lewy body type (SDLT). Psychol Med 22: 911–922

McKeith IG, Fairbairn AF, Perry R, Thompson P, Perry EK (1992b) Neuroleptic sensitivity in patients with senile dementia of Lewy body type. Br Med J 305: 673–678

McKeith IG, Fairbairn AF, Bothwell RA, Moore PB, Ferrier IN, Thompson P, Perry RH (1994) An evaluation of the predictive validity and inter-rater reliability of clinical diagnostic criteria for senile dementia of Lewy body type. Neurology 44: 872–877

McKeith IG, Fairbairn AF, Perry RH, Thompson P (1994) The clinical diagnosis and misdiagnosis of senile dementia of Lewy body type (SDLT) — do the diagnostic systems for dementia need to be revised? Br J Psychiatry 165: 324–332

Perry EK (1997) Cholinergic phytochemicals: from magic to medicine. Aging & Mental Health 1: 23–32

Perry EK, Perry RH (1995) Acetylcholine and hallucinations: disease-related compared to drug-induced alterations in human consciousness. Brain Cogn

Perry EK, Perry RH, Gibson PH, Blessed G, Tomlinson BE (1977) A cholinergic connection between normal ageing and senile dementia of the human hippocampus. Neurosci Lett 6: 85–89

Perry EK, Marshall E, Kerwin JM, Smith CJ, Jabeen S, Cheng AV, Perry RH (1990b) Evidence of a monoaminergic: cholinergic imbalance related to visual hallucinations in Lewy body dementia. J Neurochem 55(4): 1454–1456

Perry EK, Smith CJ, Court JA, Perry RH (1990c) Cholinergic nicotinic and muscarinic receptors in dementia of Alzheimer, Parkinson and Lewy body types. J Neural Transm [PD-Sect] 2: 149–158

Perry EK, Irving D, Kerwin JM, McKeith IG, Thompson P, Collerton D, Fairbairn AF, Ince PG, Morris CM, Cheng AV, Perry RH (1993) Cholinergic transmitter and neurotrophic activities in Lewy body dementia: similarity to Parkinson's and distinction from Alzheimer's disease. Alz Dis Assoc Disord 7(2): 69–79

Perry EK, Haroutunian V, Davis KL, Levy R, Lantos P, Eagger S, Honavar M, Dean A, Griffiths M, McKeith IG, Perry RH (1994) Neocortical cholinergic activities differentiate Lewy body dementia from classical Alzheimer's disease. NeuroReport 5: 747–749

Perry RH, Irving D, Blessed G, Perry EK, Fairbairn AF (1989) Clinically and neuropathologically distinct form of dementia in the elderly. Lancet i: 166

Perry RH, Irving D, Blessed G, Fairbairn AF, Perry EK (1990a) A clinically and neuropathologically distinct form of Lewy body dementia in the elderly. J Neurol Sci 95: 119–139

Piggott MA, Perry EK, McKeith IG, Marshall E, Perry RH (1994) Dopamine D_2 receptors in demented patients with severe neuroleptic sensitivity. Lancet 343: 1044–1045

Sahgal A, Galloway PG, McKeith IG, Edwardson JA, Lloyd S (1992a) A comparative study of attentional deficits in senile dementia of Alzheimer and Lewy body types. Dementia 3: 350–354

Sawle GV (1995) Living neurochemistry: the dopamine system. The Neurosciences 7: 173–177

Shergill S, Mullan E, D'Ath P, Katona C (1994) What is the clinical prevalence of Lewy body dementia? Int J Geriatr Psychiatry 9: 907–912

Wilcock GK, Scott MI (1994) Tacrine for senile dementia of Alzheimer's or Lewy body type. Lancet 344: 544

Woodard (1962) Concentric hyaline inclusion body formation in mental disease: analysis of twenty seven cases. J Neuropathol Exp Neurol 21: 442–449

Zorn SH, Jones SB, Ward KM, Liston DR (1994) Clozapine is a potent and selective muscarinic m4 receptor agonist. Eur J Mol Pharmacol 264: R1–R2

Authors' address: Dr. R. H. Perry, Department of Neuropathology, Newcastle General Hospital, Westgate Road, Newcastle upon Tyne, NE4 6BE, United Kingdom

Pathological diagnostic criteria for dementia associated with cortical Lewy bodies: review and proposal for a descriptive approach

J. Lowe[1] and **D. Dickson**[2]

[1] Department of Clinical Laboratory Sciences, Division of Pathology,
University of Nottingham Medical School, Queen's Medical Centre,
Nottingham, United Kingdom
[2] Neuropathology, Albert Einstein College of Medicine, New York, NY, U.S.A.

Summary. In recent years dementia histologically characterised by the presence of cortical Lewy bodies has been increasingly recognised. There is now need for a scheme for an internationally acceptable scheme for pathological diagnosis and classification so that clinical, pathological and molecular features of disease can be correlated. Recent observations made by different groups in large patient series have used slightly different pathological criteria resulting in at least seven different diagnostic terms. In some patients the only cortical pathology is the presence of Lewy bodies, while in the majority of patients there are coexisting pathological changes which either overlap with those seen in Alzheimer's disease (AD). Cortical Lewy bodies can also be present in patients who do not have any obvious cognitive abnormality. A problem with equating studies from different groups is that different criteria have been used to define AD, so that establishing the relevance of cortical Lewy bodies themselves to cognitive decline and separating this from the contribution which may be related to the AD pathology is problematic. The lesions which appear to be of most relevance to potential cognitive decline in DLB are cortical Lewy bodies, Lewy-related neurites, senile plaques, neurofibrillary tangles, neuronal and synaptic loss, spongiform change, and cortical cholinergic deficits. It is possible to operationally classify patients with cognitive decline and cortical Lewy bodies into three main groups, Cortical Lewy body disease, Cortical Lewy body disease with plaques, and Cortical Lewy body disease with plaques and tangles. There are frequent cases which overlap these groups making operational classification difficult in practice. A descriptive classification, in which the severity of different pathological changes is rated, is easy to use in practice. As new molecular risk factors for AD or DLB are revealed they will need to be related to morphological and clinical features. A descriptive diagnostic assessment for DLB will facilitate such studies and makes no judgements as to what these relationships will be.

Introduction

In recent years dementia histologically characterised by the presence of cortical Lewy bodies has been increasingly recognised and linked to an apparently distinct clinical phenotype (Kosaka et al., 1984; Byrne et al., 1989; Crystal et al., 1990; Hansen et al., 1990; Perry et al., 1990; Lennox, 1992; McKeith et al., 1992, 1994). As in other forms of neurodegenerative disease, progress towards treatment will depend upon insights into the molecular mechanisms underlying development of structural lesions in the brain. However, to attain this goal a scheme for pathological diagnosis and classification of dementia associated with cortical Lewy bodies needs to be evolved which will be internationally acceptable and allow comparison of results from different workers.

Present nomenclature and clinical correlates

Pathological descriptions of dementia associated with cortical Lewy bodies (DLB) were originally made as isolated case reports mainly coming from Japan. In Western countries it was around 1986 that cases were "discovered" in clinicopathological series (Clark et al., 1986; Eggertson and Sima, 1986; Byrne et al., 1987; Dickson et al., 1987; Gibb et al., 1987; Burkhardt et al., 1988).

The observations made by different groups are associated with slightly different terminology.

— Diffuse Lewy body disease (Lennox et al., 1989a; Crystal et al., 1990; Kosaka, 1990)
— Dementia associated with cortical Lewy bodies (Byrne et al., 1991)
— Lewy body dementia (Sima et al., 1986)
— Diffuse cortical Lewy body disease (Gibb et al., 1987)
— Cortical Lewy body dementia (Gibb et al., 1989)
— Senile dementia of Lewy body type (Perry et al., 1990)
— Lewy body variant of Alzheimer's disease (Hansen et al., 1990)
— Lewy body disease (Lippa et al., 1994)

Differences in description of patients derived from different centres can be largely related to referral bias and on the operational definition of Alzheimer's disease being used for pathological diagnosis. It is clear that in some cases the only cortical histological abnormality is the presence of Lewy bodies, while in the majority of cases there are coexisting changes which either overlap with or are identical with those seen in Alzheimer's disease (AD).

Dementia associated with cortical Lewy bodies has been divided into two types: a common form associated with changes of AD in the cerebral cortex, and a pure form with none or few changes of AD. While this suggests that cortical Lewy bodies alone can cause cortical dementia (Kosaka, 1993), there remains great uncertainty as to the contribution cortical Lewy bodies make to cognitive decline in patients who also have AD changes (de Vos et al., 1995). It has become apparent that cortical Lewy body pathology may be related to the presence of dementia in patients who have many cortical plaques but few or no tangles (Hansen et al., 1993). Other evidence suggests that cortical Lewy bodies can be present in patients who do not have any obvious cognitive abnormality, being seen in low density in most patients with idiopathic Parkinson's disease (Hughes et al., 1993).

A major problem with equating studies from different centres is that different criteria have been used to define AD, so that establishing the relevance of cortical Lewy bodies themselves to cognitive decline and separating this from the contribution which may be related to the AD pathology is problematic. Now that molecular risk factors are emerging for both Alzheimer's disease and Lewy body Parkinson's disease the differences in pathological definition of dementia associated with cortical Lewy bodies also affects understanding of its likely molecular pathogenesis. There is pressing need for an internationally agreed standardised method for pathological assessment in this pattern of dementia.

Clinically, the spectrum of diseases associated with cortical Lewy bodies is still being defined.

— Parkinsonism with no cognitive decline (Hughes et al., 1993)
— Cortical dementia syndrome (Hansen et al., 1990)
— Parkinsonism associated with dementia (Byrne et al., 1989)
— Isolated dysphagia (Jackson et al., 1995)
— Progressive supranuclear palsy (Mori et al., 1986; Fearnley et al., 1991; de Bruin et al., 1992)
— Dystonia (Kulisevsky et al., 1988; Olsson et al., 1992; Mark et al., 1994)
— Motor neuron disorder (Hainfellner et al., 1995)

Dementia associated with cortical Lewy bodies should therefore be regarded as only one member of the set of clinical disorders characterised by the presence of cortical Lewy bodies.

What pathological changes are important?

In defining pathological diagnostic criteria for dementia associated with cortical Lewy bodies the structural lesions which appear relevant can be defined from review of previous pathological descriptions.

— Lewy bodies can be readily identified in conventional histological preparations and divided into classical and cortical types (Pollanen et al., 1993; Lowe, 1994). Cortical Lewy bodies can be identified using immunohistochemical techniques, especially anti-ubiquitin (Lennox et al., 1989b), however this is not essential for their identification. Unless care is taken, other inclusions detected with anti-ubiquitin can be confused with cortical Lewy bodies (Lowe, 1994)
— Lewy-associated neurites are not visible using conventional stains and are detected with anti-ubiquitin. They appear to be closely associated with the presence of Lewy body pathology in the brain but are not specifically associated with dementia (Dickson et al., 1991, 1994; Braak et al., 1994; Gai et al., 1995; Kim et al., 1995; Pellise et al., 1996)

— Senile plaques are present in most, but not all, cases of dementia associated with cortical Lewy bodies. These have been found to have characteristics of both diffuse and neuritic plaques, diffuse plaques lacking associated Tau-immunoreactive neurites being the most common type (Dickson et al., 1989; Gentleman et al., 1992)

— Tau pathology is present in many cases of dementia associated with cortical Lewy bodies in the form of neurofibrillary tangles, plaque neurites and neuropil threads. Studies have suggested that the extent of cortical involvement with Tau-pathology is less marked than in patients with "pure" Alzheimer's disease (Hansen et al., 1991; Ince et al., 1991; Harrington et al., 1994b; Lippa et al., 1994)

— Neuronal loss is seen in dementia associated with cortical Lewy bodies but is generally less marked than in comparable cases of AD (Lippa et al., 1994)

— Loss of immunoreactivity for synaptic proteins has been related to cognitive decline in both LBD and AD (Zhan et al., 1993; Wakabayashi et al., 1994)

— Spongiform change in a restricted distribution in the temporal lobe is a characteristic but poorly understood feature of dementia associated with cortical Lewy bodies (Hansen et al., 1989; Armstrong et al., 1991)

— Neocortical cholinergic activity is greatly reduced in many cases of LBD and it has been suggested that this is distinctive and responsible for clinical improvement when treated with cholinergic therapy (Perry et al., 1994)

— Amyloid angiopathy has been related to the presence of Alzheimer's disease changes in the presence of cortical Lewy bodies (Wu et al., 1992).

Quantitative diagnostic criteria

While it is possible to make a quantitative assessment of the density of cortical Lewy bodies and other pathological lesions it seems unwise at present to use any quantitative threshold to define disease. Although some studies have shown a correlation between density of cortical Lewy bodies and dementia there are at present no firm data on which to state a density of cortical Lewy bodies above which patients are always demented or below which they are always cognitively normal (Lennox et al., 1989a; Samuel et al., 1996). The regional distribution of cortical Lewy bodies is, however, likely to be related to the presence or absence of clinical dementia. It is our practical experience that in Parkinson's disease without clinical cognitive abnormality cortical Lewy bodies are limited to limbic areas. In contrast, the presence of more than rare cortical Lewy bodies in frontal and parietal association neocortices is almost invariably associated with clinical cognitive abnormality. We would therefore argue that the regional distribution of cortical Lewy bodies is of more diagnostic importance than their density. There is presently no evidence to suggest that patients who have dementia associated with cortical Lewy bodies progress from the normal state, through limbic-restricted disease to a more widespread disease. This is in contrast with the proposed stage development of Alzheimer's disease (Braak and Braak, 1991).

Operational diagnostic criteria

Based on descriptive features, operational criteria can be used to assign cases into different diagnostic groups. Such divisions are easy to construct and can include clinical features for purposes of diagnostic grouping. Based on previous pathological studies it is possible to delineate several pathological subsets of dementia associated with cortical Lewy bodies.

Cortical Lewy body disease

Neurodegenerative process in which Lewy body pathology principally affects the cortex in the absence of neurofibrillary tangles/Tau-pathology in neocortex. In addition to affecting elderly patients cases may be seen in younger individuals (Yoshimura et al., 1988; Kosaka, 1993). Cortical Lewy bodies are typically present in temporal, frontal and parietal neocortex as well as in limbic regions. Amyloid plaques are absent. Lewy bodies are almost always also seen in brain stem nuclei (substantia nigra and locus coeruleus). Very rare cases may be been without significant brain stem involvement (Kosaka et al., 1994). The main clinical correlate is with a cortical dementia syndrome, usually associated with an extrapyramidal movement disorder, sometimes associated with autonomic failure.

Cortical Lewy body disease with plaques

Neurodegenerative process usually seen in the elderly in which Lewy body pathology principally affects the cortex in the absence of neurofibrillary tangles/Tau-pathology in neocortex. Cortical Lewy bodies are typically present in temporal, frontal and parietal neocortex as well as in limbic regions. Amyloid plaques are present in the absence of Tau-reactive neurites, although neurites immunoreactive for ubiquitin may be seen. Lewy bodies are almost always also seen in brain stem nuclei (substantia nigra and locus coeruleus). The main clinical correlate is with a cortical dementia syndrome, often associated with an extrapyramidal movement disorder, sometimes associated with autonomic failure. In view of the fact that similar plaques can be seen in cognitively normal individuals it is possible that they are not related to development of dementia.

Cortical Lewy body disease with plaques and tangles

Neurodegenerative process usually seen in the elderly in which there is cortical Lewy body pathology associated with neurofibrillary tangles/

Tau-pathology in neocortex. Cortical Lewy bodies are typically present in temporal, frontal and parietal neocortex as well as in limbic regions. Amyloid plaques have associated Tau pathology. The plaque and tangle pathology are alone sufficient to make a diagnosis of Alzheimer's disease (CERAD) (Mirra et al., 1991). Lewy bodies are almost always also seen in brain stem nuclei (substantia nigra and locus coeruleus). The main clinical correlate is with a cortical dementia syndrome, sometimes in the presence of extrapyramidal movement disorder.

Brain stem Lewy body disease

Neurodegenerative process in which Lewy body pathology principally affects the substantia nigra, locus coeruleus, dorsal vagal nucleus, and nucleus basalis of Meynert. Cortical Lewy bodies are usually present in small number and are restricted to limbic regions. The main clinical correlate is with Parkinson's disease, or primary dysphagia (symptomatic Lewy body disease). This pathology may also be seen without clinically significant manifestations (incidental Lewy body disease).

Descriptive diagnostic criteria

The main problems with using operational criteria is that there is frequent overlap between groups and changes which may be attributable to age may confound placement of a case in any particular group.

At present we favour using a descriptive approach to pathological classification for dementia associated with Lewy body pathology.

— State Lewy body/Lewy neurite distribution
  Rate density
  Rate cell loss
— State presence of Alzheimer-associated pathology

Rate the degree of abnormality using one of the widely used and accepted methods for example (Braak and Braak, 1991; Mirra et al., 1991; Bancher et al., 1993; Gearing et al., 1995).

— State associated vascular-related pathology using one of the widely used and accepted methods (Mirra et al., 1991)
— State associated clinical features

This approach will allow correlation between morphological changes and emerging molecular risk factors for Lewy body pathology. An appropriate rating scheme for cell loss, and density of Lewy bodies and Lewy neurites which is reproducible between laboratories remains to be defined.

Relevance of a descriptive classification to future understanding

Lewy body disease will probably be explained by a network of interacting susceptibility factors. It is well recognised that there are genetically determined predispositions for some individuals to develop AD and it is not unreasonable to imagine how this expression could be modulated by any associated predisposing factors to develop Lewy bodies. For example it is apparent that patients with trisomy 21 as well as patients with mutations in APP can also have cortical Lewy body disease (Bodhireddy et al., 1994; Hardy, 1994; Lantos et al., 1994). A high frequency of Apo E4 allele also appears to be related to the presence of Alzheimer-type pathology in cases of dementia associated with cortical Lewy bodies (Benjamin et al., 1994; Betard et al., 1994; Galasko et al., 1994; Hansen et al., 1994; Hardy et al., 1994; Harrington et al., 1994a; St Clair et al., 1994; Lippa et al., 1995).

It is very likely that the factors which are discovered that predispose to hereditary Lewy body Parkinson's disease will also be of relevance to DLB (Duvoisin and Johnson, 1992; Golbe, 1993). For example the cytochrome P450 CYP2D6–debrisoquine 4-hydroxylase mutant B allele, a susceptibility gene for PD, is more prevalent in the Lewy body variant of AD than in pure AD or non-AD without Lewy bodies (Saitoh et al., 1995). Recent neuropathological review of patients with documented familial dementia associated with parkinsonism has revealed that the pathology in many cases is that of cortical Lewy body disease (Mark et al., 1996).

As new molecular risk factors for AD or DLB become apparent they will need to be related to specific morphological and clinical features. A descriptive diagnostic assessment for cases of dementia associated with cortical Lewy bodies will facilitate such studies and makes no judgements as to what these relationships will be.

References

Armstrong TP, Hansen LA, Salmon DP, Masliah E, et al (1991) Rapidly progressive dementia in a patient with the Lewy body variant of Alzheimer's disease. Neurology 41(8): 1178–1180

Bancher C, Braak H, Fischer P, Jellinger KA (1993) Neuropathological staging of Alzheimer lesions and intellectual status in Alzheimer's and Parkinson's disease patients. Neurosci Lett 162(1–2): 179–182

Benjamin R, Leake A, Edwardson JA, McKeith IG, et al (1994) Apolipoprotein E genes in Lewy body and Parkinson's disease. Lancet 343(8912): 1565

Betard C, Robitaille Y, Gee M, Tiberghien D, et al (1994) Apo E allele frequencies in Alzheimer's disease, Lewy body dementia, Alzheimer's disease with cerebrovascular disease and vascular dementia. Neuroreport 5(15): 1893–1896

Bodhireddy S, Dickson DW, Mattiace L, Weidenheim KM (1994) A case of Down's syndrome with diffuse Lewy body disease and Alzheimer's disease. Neurology 44(1): 159–161

Braak H, Braak E (1991) Neuropathological staging of Alzheimer-related changes. Acta Neuropathol (Berl) 82(4): 239–259

Braak H, Braak E, Yilmazer D, de Vos RA, et al (1994) Amygdala pathology in Parkinson's disease. Acta Neuropathol (Berl) 88(6): 493–500

Burkhardt CR, Filley CM, Kleinschmidt-De Masters BK, de la Monte S, et al (1988) Diffuse Lewy body disease and progressive dementia. Neurology 38(10): 1520–1528

Byrne EJ, Lowe J, Godwin-Austen RB, Arie T, et al (1987) Dementia and Parkinson's disease associated with diffuse cortical Lewy bodies. Lancet i (8531): 501

Byrne EJ, Lennox G, Lowe J, Godwin-Austen RB (1989) Diffuse Lewy body disease: clinical features in 15 cases. J Neurol Neurosurg Psychiatry 52(6): 709–717

Byrne E, Lennox G, Godwin-Austen R, Mayer R, et al (1991) Diagnostic criteria for dementia associated with cortical Lewy bodies. Dementia 2: 283–284

Clark AW, White CLd, Manz HJ, Parhad IM, et al (1986) Primary degenerative dementia without Alzheimer pathology. Can J Neurol Sci 13 [4 Suppl]: 462–470

Crystal HA, Dickson DW, Lizardi JE, Davies P, et al (1990) Antemortem diagnosis of diffuse Lewy body disease. Neurology 40(10): 1523–1528

de Bruin VM, Lees AJ, Daniel SE (1992) Diffuse Lewy body disease presenting with supranuclear gaze palsy, parkinsonism, and dementia: a case report. Mov Disord 7(4): 355–358

de Vos RA, Jansen EN, Stam FC, Ravid R, et al (1995) "Lewy body disease": clinico-pathological correlations in 18 consecutive cases of Parkinson's disease with and without dementia. Clin Neurol Neurosurg 97(1): 13–22

Dickson DW, Davies P, Mayeux R, Crystal H, et al (1987) Diffuse Lewy body disease. Neuropathological and biochemical studies of six patients. Acta Neuropathol (Berl) 75(1): 8–15

Dickson DW, Crystal H, Mattiace LA, Kress Y, et al (1989) Diffuse Lewy body disease: light and electron microscopic immunocytochemistry of senile plaques. Acta Neuropathol (Berl) 78(6): 572–584

Dickson DW, Ruan D, Crystal H, Mark MH, et al (1991) Hippocampal degeneration differentiates diffuse Lewy body disease (DLBD) from Alzheimer's disease: light and electron microscopic immunocytochemistry of CA2-3 neurites specific to DLBD. Neurology 41(9): 1402–1409

Dickson DW, Schmidt ML, Lee VM, Zhao ML, et al (1994) Immunoreactivity profile of hippocampal CA2/3 neurites in diffuse Lewy body disease. Acta Neuropathol (Berl) 87(3): 269–276

Duvoisin RC, Johnson WG (1992) Hereditary Lewy-body parkinsonism and evidence for a genetic etiology of Parkinson's disease. Brain Pathol 2(4): 309–320

Eggertson DE, Sima AA (1986) Dementia with cerebral Lewy bodies. A mesocortical dopaminergic defect? Arch Neurol 43(5): 524–527

Fearnley JM, Revesz T, Brooks DJ, Frackowiak RS, et al (1991) Diffuse Lewy body disease presenting with a supranuclear gaze palsy. J Neurol Neurosurg Psychiatry 54(2): 159–161

Gai WP, Blessing WW, Blumbergs PC (1995) Ubiquitin-positive degenerating neurites in the brainstem in Parkinson's disease. Brain 118(6): 1447–1459

Galasko D, Saitoh T, Xia Y, Thal LJ, et al (1994) The apolipoprotein E allele epsilon 4 is over-represented in patients with the Lewy body variant of Alzheimer's disease. Neurology 44(10): 1950–1951

Gearing M, Mirra SS, Hedreen JC, Sumi SM, et al (1995) The Consortium to Establish a Registry for Alzheimer's Disease (CERAD), part X. Neuropathology confirmation of the clinical diagnosis of Alzheimer's disease. Neurology 45(3 Pt 1): 461–466

Gentleman SM, Williams B, Royston MC, Jagoe R, et al (1992) Quantification of beta A4 protein deposition in the medial temporal lobe: a comparison of Alzheimer's disease and senile dementia of the Lewy body type. Neurosci Lett 142(1): 9–12

Gibb WR, Esiri MM, Lees AJ (1987) Clinical and pathological features of diffuse cortical Lewy body disease (Lewy body dementia). Brain 110(Pt 5): 1131–1153

Gibb WR, Luthert PJ, Janota I, Lantos PL (1989) Cortical Lewy body dementia: clinical features and classification. J Neurol Neurosurg Psychiatry 52(2): 185–192

Golbe LI (1993) The genetics of Parkinson's disease. Rev Neurosci 4(1): 1–16

Hainfellner JA, Pilz P, Lassmann H, Ladurner G, et al (1995) Diffuse Lewy body disease as substrate of primary lateral sclerosis. J Neurol 242(2): 59–63

Hansen LA, Masliah E, Terry RD, Mirra SS (1989) A neuropathological subset of Alzheimer's disease with concomitant Lewy body disease and spongiform change. Acta Neuropathol (Berl) 78(2): 194–201

Hansen L, Salmon D, Galasko D, Masliah E, et al (1990) The Lewy body variant of Alzheimer's disease: a clinical and pathologic entity. Neurology 40(1): 1–8

Hansen LA, Masliah E, Quijada-Fawcett S, Rexin D (1991) Entorhinal neurofibrillary tangles in Alzheimer disease with Lewy bodies. Neurosci Lett 129(2): 269–272

Hansen LA, Masliah E, Galasko D, Terry RD (1993) Plaque-only Alzheimer disease is usually the lewy body variant, and vice versa. J Neuropathol Exp Neurol 52(6): 648–654

Hansen LA, Galasko D, Samuel W, Xia Y, et al (1994) Apolipoprotein-E epsilon-4 is associated with increased neurofibrillary pathology in the Lewy body variant of Alzheimer's disease. Neurosci Lett 182(1): 63–65

Hardy J (1994) Lewy bodies in Alzheimer's disease in which the primary lesion is a mutation in the amyloid precursor protein. Neurosci Lett 180 (2): 290–291

Hardy J, Crook R, Prihar G, Roberts G, et al (1994) Senile dementia of the Lewy body type has an apolipoprotein E epsilon 4 allele frequency intermediate between controls and Alzheimer's disease. Neurosci Lett 182(1): 1–2

Harrington CR, Louwagie J, Rossau R, Vanmechelen E, et al (1994a) Influence of apolipoprotein E genotype on senile dementia of the Alzheimer and Lewy body types. Significance for etiological theories of Alzheimer's disease. Am J Pathol 145(6): 1472–1484

Harrington CR, Perry RH, Perry EK, Hurt J, et al (1994b) Senile dementia of Lewy body type and Alzheimer type are biochemically distinct in terms of paired helical filaments and hyperphosphorylated tau protein. Dementia 5(5): 215–228

Hughes AJ, Daniel SE, Blankson S, Lees AJ (1993) A clinicopathologic study of 100 cases of Parkinson's disease. Arch Neurol 50(2): 140–148

Ince P, Irving D, MacArthur F, Perry RH (1991) Quantitative neuropathological study of Alzheimer-type pathology in the hippocampus: comparison of senile dementia of Alzheimer type, senile dementia of Lewy body type, Parkinson's disease and non-demented elderly control patients. J Neurol Sci 106(2): 142–152

Jackson M, Lennox G, Balsitis M, Lowe J (1995) Lewy body dysphagia. J Neurol Neurosurg Psychiatry 58(6): 756–758

Kim H, Gearing M, Mirra SS (1995) Ubiquitin-positive CA2/3 neurites in hippocampus coexist with cortical Lewy bodies. Neurology 45(9): 1768–1770

Kosaka K (1990) Diffuse Lewy body disease in Japan. J Neurol 237(3): 197–204

Kosaka K (1993) Dementia and neuropathology in Lewy body disease. Adv Neurol 60: 456–463

Kosaka K, Yoshimura M, Ikeda K, Budka H (1984) Diffuse type of Lewy body disease: progressive dementia with abundant cortical Lewy bodies and senile changes of varying degree — A new disease? Clin Neuropathol 3(5): 185–192

Kosaka K, Iseki E, Odawara T (1994) Cerebral type of Lewy body disease — a case report. Neuropathology 16: 72–75

Kulisevsky J, Marti MJ, Ferrer I, Tolosa E (1988) Meige syndrome: neuropathology of a case. Mov Disord 3(2): 170–175

Lantos PL, Ovenstone IM, Johnson J, Clelland CA, et al (1994) Lewy bodies in the brain of two members of a family with the 717 (Val to Ile) mutation of the amyloid precursor protein gene. Neurosci Lett 172(1–2): 77–79

Lennox G (1992) Lewy body dementia. Baillières Clin Neurol 1(3): 653–676

Lennox G, Lowe J, Landon M, Byrne EJ, et al (1989a) Diffuse Lewy body disease: correlative neuropathology using anti-ubiquitin immunocytochemistry. J Neurol Neurosurg Psychiatry 52(11): 1236–1247

Lennox G, Lowe J, Morrell K, Landon M, et al (1989b) Anti-ubiquitin immunocytochemistry is more sensitive than conventional techniques in the detection of diffuse Lewy body disease. J Neurol Neurosurg Psychiatry 52(1): 67–71

Lippa CF, Smith TW, Swearer JM (1994) Alzheimer's disease and Lewy body disease: a comparative clinicopathological study [published erratum appears in Ann Neurol (1994) 35(3):380]. Ann Neurol 35(1): 81–88

Lippa CF, Smith TW, Saunders AM, Crook R, et al (1995) Apolipoprotein E genotype and Lewy body disease. Neurology 45(1): 97–103

Lowe J (1994) Lewy bodies. In: Calne D (ed) Neurodegenerative diseases. Saunders, Philadelphia, pp 51–69

Mark MH, Sage JI, Dickson DW, Heikkila RE, et al (1994) Meige syndrome in the spectrum of Lewy body disease. Neurology 44(8): 1432–1436

Mark M, Dickson D, Sage J, Duvoisin R (1996) The clinicopathologic spectrum of Lewy body disease. Adv Neurol 69: 315–318

McKeith IG, Perry RH, Fairbairn AF, Jabeen S, et al (1992) Operational criteria for senile dementia of Lewy body type (SDLT). Psychol Med 22(4): 911–922

McKeith IG, Fairbairn AF, Bothwell RA, Moore PB, et al (1994) An evaluation of the predictive validity and inter-rater reliability of clinical diagnostic criteria for senile dementia of Lewy body type. Neurology 44(5): 872–877

Mirra SS, Heyman A, McKeel D, Sumi SM, et al (1991) The Consortium to Establish a Registry for Alzheimer's Disease (CERAD), part II. Standardization of the neuropathologic assessment of Alzheimer's disease. Neurology 41(4): 479–486

Mori H, Yoshimura M, Tomonaga M, Yamanouchi H (1986) Progressive supranuclear palsy with Lewy bodies. Acta Neuropathol (Berl) 71(3–4): 344–346

Olsson JE, Brunk U, Lindvall B, Eeg-Olofsson O (1992) Dopa-responsive dystonia with depigmentation of the substantia nigra and formation of Lewy bodies. J Neurol Sci 112(1–2): 90–95

Pellise A, Roig C, BarraquerBordas L, Ferrer I (1996) Abnormal, ubiquitinated cortical neurites in patients with diffuse Lewy body disease. Neurosci Lett 206(2–3): 85–88

Perry RH, Irving D, Blessed G, Fairbairn A, et al (1990) Senile dementia of Lewy body type. A clinically and neuropathologically distinct form of Lewy body dementia in the elderly. J Neurol Sci 95(2): 119–139

Perry EK, Haroutunian V, Davis KL, Levy R, et al (1994) Neocortical cholinergic activities differentiate Lewy body dementia from classical Alzheimer's disease. Neuroreport 5(7): 747–749

Pollanen MS, Dickson DW, Bergeron C (1993) Pathology and biology of the Lewy body. J Neuropathol Exp Neurol 52(3): 183–191

Saitoh T, Xia Y, Chen X, Masliah E, et al (1995) The CYP2D6B mutant allele is over-represented in the Lewy body variant of Alzheimer's disease. Ann Neurol 37(1): 110–112

Samuel W, Galasko D, Masliah E, Hansen LA (1996) Neocortical Lewy body counts correlate with dementia in the Lewy body variant of Alzheimer's disease. J Neuropathol Exp Neurol 55(1): 44–52

Sima AA, Clark AW, Sternberger NA, Sternberger LA (1986) Lewy body dementia without Alzheimer changes. Can J Neurol Sci 13 [4 Suppl]: 490–497

St Clair D, Norman J, Perry R, Yates C, et al (1994) Apolipoprotein E epsilon 4 allele frequency in patients with Lewy body dementia, Alzheimer's disease and age-matched controls. Neurosci Lett 176(1): 45–46

Wakabayashi K, Honer WG, Masliah E (1994) Synapse alterations in the hippocampal-entorhinal formation in Alzheimer's disease with and without Lewy body disease. Brain Res 667(1): 24–32

Wu E, Lipton RB, Dickson DW (1992) Amyloid angiopathy in diffuse Lewy body disease. Neurology 42(11): 2131–2135

Yoshimura N, Yoshimura I, Asada M, Hayashi S, et al (1988) Juvenile Parkinson's disease with widespread Lewy bodies in the brain. Acta Neuropathol (Berl) 77(2): 213–218

Zhan SS, Beyreuther K, Schmitt HP (1993) Quantitative assessment of the synaptophysin immuno-reactivity of the cortical neuropil in various neurodegenerative disorders with dementia. Dementia 4(2): 66–74

Authors' address: Prof. J. Lowe, Department of Clinical Laboratory Sciences, Division of Pathology, University of Nottingham Medical School, Queens Medical Centre, Nottingham NG7 2UH, United Kingdom

Pathology of familial Alzheimer's disease with Lewy bodies

T. Revesz[1], J. L. McLaughlin[2], M. N. Rossor[3], and P. L. Lantos[4]

[1]Department of Neuropathology, Institute of Neurology,
[2]Royal Free Hospital, [4]Institute of Psychiatry, and [3]Department of Clinical Neurology,
Institute of Neurology, London, United Kingdom

Summary. The neuropathological findings of three cases from two pedigrees with early onset familial Alzheimer's disease (FAD) are reported. Affected members of the first family, including cases 1 and 2 reported here, are known to have 717 valine to isoleucine mutation of the amyloid precursor protein (APP) gene, while the genetic background of the disease has not been clarified yet in the second family. In all three cases, in addition to the classical histological findings associated with Alzheimer's disease (AD), both nigral and cortical Lewy bodies (LBs) occurred.

The association of LBs with AD type pathology, which may be observed in both sporadic and familial AD, raises important nosological issues. These include a possible overlap between AD and other neurodegenerative conditions presenting primarily with LBs. In this respect the clinically and neuropathologically distinct disease entity described under the terms of "senile dementia of the Lewy body type" and "Lewy body variant of AD" may be especially important. The occurrence of LBs in association with severe AD-type histological changes in sporadic and especially in some familial AD cases indicate that these inclusions may be another expression of the altered cytoskeleton in AD.

Introduction

AD, which affects about 400,000 people in the United Kingdom alone, is the commonest cause of dementia throughout the Western world (Rossor, 1993a). Although AD primarily is a sporadic condition, at least some of its forms have a genetic aetiology with a clear autosomal dominant pattern of inheritance in around 5–10% of cases (Rossor, 1993b). The process of clarifying the genetic basis of FAD was accelerated by the discovery of different mutations in the APP gene located on chromosome 21 (Goate et al., 1991; Chartier-Harlin et al., 1991; Murrell et al., 1991; Mullan et al., 1992). However subsequent studies have shown that the APP gene mutations are relatively rare events, occurring in less than 10% of all early-onset familial cases (Mullan and Crawford, 1993; Bird, 1994), and that the majority of early onset familial cases are due to mutations in a recently cloned gene on chromosome 14 (Sherrington et al., 1995). Furthermore a third autosomal dominant locus at 1q31–42 has been demonstrated in Volga German kindreds with early-onset disease (Levy-Lahad et al., 1995a,b). In addition to these major gene abnormalities, the ε4 allele of the apolipoprotein E gene has been found to modify the risk of developing some forms of AD including late-onset FAD, probably by lowering the age of onset in these cases (Corder et al., 1993).

It is generally agreed that the neuropathological changes of FAD are indistinguishable from those of sporadic AD (Lantos, 1992). In both the sporadic and familial forms the occasional coexistence of brainstem and cortical LBs, which are hallmarks of Parkinson's disease and "diffuse Lewy body disease', with otherwise typical histological changes of AD may be explained by collision of two diverse conditions (Gibb et al., 1989). However clinico-pathological entities in which the combination of the two pathologies appears to be a constant finding have been described under the terms of "senile dementia of the Lewy body type" and the "Lewy body variant of AD" (Perry et al., 1990; Hansen et al., 1989, 1990).

Here we present three early-onset FAD cases from two families, all three showing both brainstem and cortical LBs, in addition to typical histological changes of AD. Cases 1 and 2 (family 1) are from a large pedigree with 717 valine to isoleucine mutation, and both have been reported previously (Lantos et al., 1992, 1994). Case 3 is from an unrelated family (family 2) with an as yet unexplored genetic background. One of the interesting features of this latter family's disease is that its affected members develop dementia in their late twenties and die around their mid-thirties (McLaughlin et al., 1993).

Patients and methods

Family 1

Several members of this family are known to have suffered from early onset AD (Fig. 1). Brains from two members of this family who died from the disease have been examined.

Case 1 (III-4)

At the age of 41 years this right-handed housewife sustained an unexplained cardiac arrest following hysterectomy but made full recovery from this without any cognitive impairment. At the age of 48 years she lost her sense of smell and this was investigated by an ENT surgeon but without a cause being established. Three years later at the age of 51 she developed memory impairment with a tendency to mislay items and then had difficulty with her domestic chores including cooking. She had little insight into her problems. She was referred for investigation at the age of 53 years when she was found on simple bedside testing to have problems with orientation in time and space, difficulty with drawing geometric figures and a degree of apraxia. She was also noted to be anosmic but other aspects of the neurological examination were unremarkable. The CT scan was reported as showing cerebral atrophy but other investigations were normal. A clinical diagnosis of AD was made. Over the next 2 years there was inexorable decline as she

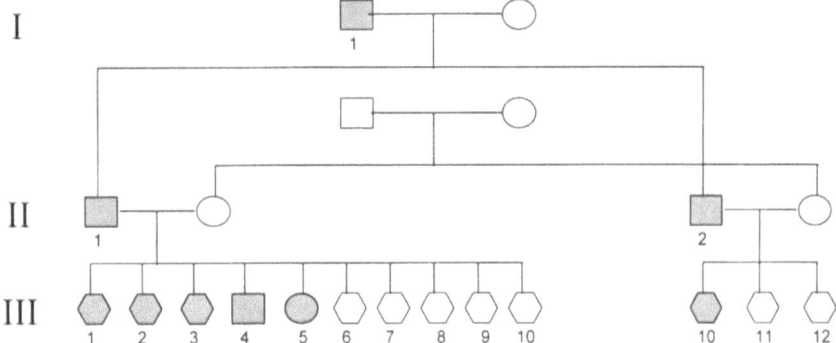

Fig. 1. Pedigree of family 1. To preserve anonymity no further information is given about other members of generation III but III4 and III5

developed increased tone and a stooped posture but this was attributed to neuroleptic medication. Late in the course of her illness she developed generalised seizures and myoclonus and died from bronchopneumonia at the age of 59 years.

Case 2 (III-5)

This left-handed milkman developed difficulties keeping his accounts at the age of 58 years and the following year began to mislay items. He was admitted for investigation when poor memory was documented but no other neurological abnormality was found. CT scan and CSF examination were normal. EEG showed some left temporal slowing. A clinical diagnosis of AD was made and he retired from his job. At the age of 64 years he was admitted for further assessment at which time he was unable to dress himself and required assistance with reading. His speech was virtually incomprehensible and he was unable to name objects although reading simple passages was preserved. The following year he sustained a brief episode of right arm weakness with drowsiness with complete resolution. However, over the next four years he became increasingly rigid and subsequently unable to walk due to a severe gait apraxia. Primitive reflexes emerged during this time together with prominent myoclonus.

Family 2

Four members of this family, including the mother (III-3), maternal aunt (III-2) and grandfather (II-1) of the patient described below (IV-2) are known to have suffered from AD. In addition the maternal great grandmother (I-1) probably had the same disease (Fig. 2). The diagnosis of AD had been confirmed by neuropathological examination in III-2, but material was no longer available for the current review.

Case 3 (IV-2)

The patient, a right-handed female was investigated at the age of 30 years with a three years' history of progressive cognitive impairment. In the past medical history she had undergone an appendectomy at the age of 12 years and treatment for tuberculosis at the age of 15 years. There had also been a brief self-limiting episode of depression following

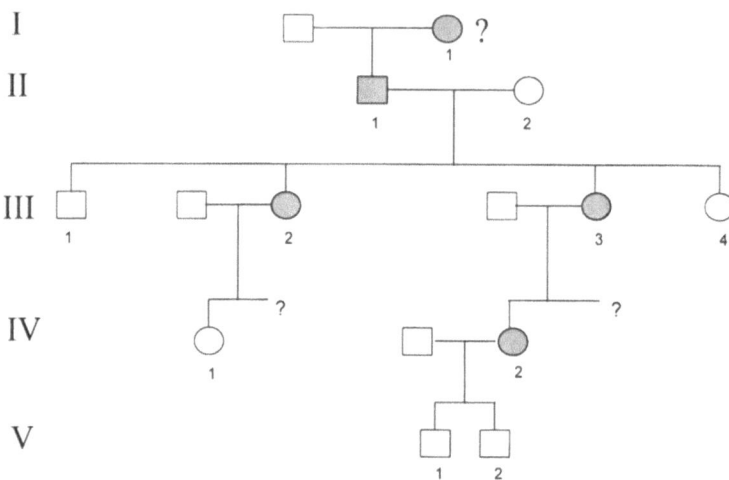

Fig. 2. Pedigree of family 2

the birth of her second child at the age of 25 years. No cognitive impairment was noted at that time.

The initial symptoms were of memory impairment and progressive difficulties with coping at work. When investigated at the age of 30 years, she obtained a full scale IQ of 74 with evidence of widespread impairment in many cognitive domains. A CT performed at that time showed widespread atrophy. She was investigated again at the age of 33 years when examination of the CSF revealed the presence of oligoclonal bands but when repeated a year later were negative. On neither admission were extrapyramidal features noted but in the last year of her illness she developed prominent myoclonus and generalised seizures requiring treatment with sodium valproate. At 34 years of age she suddenly died during a respite hospital admission.

Sampling

The brains in cases 1 and 2 (family 1) were bisected through the midline in the sagittal plane and then sliced coronally. Extensive samples were frozen while the remainder of the material was fixed in 10% buffered formalin and was later examined morphologically. In case 3 (family 2) the whole formalin fixed brain was available for neuropathological examination. In all three cases the major anatomical areas including all four lobes, deep grey nuclei, brainstem and cerebellum, and in case 3 the spinal cord were thoroughly sampled and tissue blocks were processed in paraffin wax.

Staining and immunohistochemistry

Tissue sections were cut at 7–14 μm and were stained with haematoxylin and eosin, luxol fast blue/cresyl violet and Bielschowsky's silver methods. Sections from selected blocks were also used for Gallyas impregnation. Immunohistochemistry was carried out using the ABC method with antibodies to ubiquitin, tau phosphoprotein, Aβ peptide, RT97 and GFAP. By applying a number of antibodies, shown in Table 1, the cytoskeletal pathology was analysed and compared to a sporadic case of AD in case 1.

Table 1. Immunohistochemistry for cytoskeletal proteins

		Staining of tangles and plaque neurites	
Antibody	Dilution	Case 1	Sporadic AD case
Ubiquitin	1:100	+	+
Human tau	1:1,000	+	+
Tau 1*	1:10	+	+
Phosphorylated neurofilaments:			
147	1:500	−	−
155	1:500	−	−
RS18	1:500	−	−
8D8	1:500	+	+
BF10	1:200	+	+
RT97	1:100	+	+
121S	1:500	+	+

*immunoreaction after alkaline phosphatase pretreatment

Morphometry

Quantitation of cortical neurofibrillary tangles (NFTs) and senile plaques (SPs) was carried out in all three cases. In brief the density of the NFTs and SPs (N/mm^2) was determined in the middle frontal, first temporal gyri, inferior parietal lobule and striate cortex at $\times 400$ magnification. The numbers of pigmented nerve cells of the zona compacta of the substantia nigra and of LB and NFT-bearing neurons were determined in case 3. For comparison nigral neuronal cell counts from corresponding levels were determined in an age-matched control.

The density of cortical Lewy bodies detected on haematoxylin and eosin stained sections was also determined in several cortical areas in case 1.

Results

Neuropathological findings

After fixation the complete half brain weighed 380 g (brainstem and cerebellum 69 g) in case 1, and the full brain 1,069 g (brainstem and cerebellum 167 g) in case 3. This information was not available in case 2. The cerebral hemispheres showed severe generalised and symmetrical atrophy with thinning of the gyri and widening of the sulci in cases 1 and 2 whereas the atrophy was only moderate in case 3. The enlargement of the lateral ventricles with rounding of the lateral angle was more severe in cases 1 and 2 than in case 3. The space between the hippocampus and the wall of the temporal horns was enlarged in all three cases. The major blood vessels of the circle of Willis were thin-walled in case 3, but showed only occasional atherosclerotic plaques in case 1. The basal ganglia, thalami and subthalamic nuclei were of normal size and macroscopical appearance. The substantia nigra and loci coerulei were paler than normal in all three cases, otherwise the brainstem macroscopically was normal, as was the cerebellum.

Histology

All isocortical areas, including the association cortices were invariably affected by the disease process. The molecular layer was often filled with faintly stained and frequently coalescent Aβ positive amyloid deposits in the different isocortical areas while layers II and III much more often contained mature SPs. NFTs and neuropil threads were numerous in layers II, III and also in layer V. In general the NFTs, neuropil threads and dystrophic neurites of the Aβ peptide positive SPs were readily demonstrated with the modified Bielschowsky and Gallyas silver methods (Fig. 3a). These structures including NFTs were strongly positive with antibodies to tau (Fig. 3b), while anti-ubiquitin antibodies usually stained a relatively smaller proportion of the tangles than anti-tau antibodies. The often confluent plaques of the deeper cerebral neocortical layers were rather faint on both the Bielschowsky's silver impregnation and Aβ immunostain.

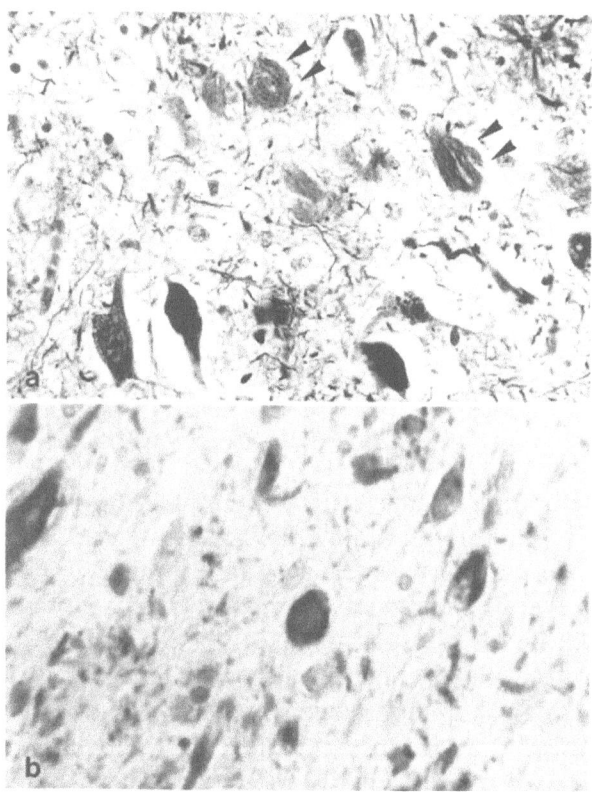

Fig. 3. Silver (**a**) and tau-positive (**b**) neurofibrillary tangles in the hippocampus many of which are of the ghost tangle type (arrowheads). **a** Bielschowsky silver impregnation. ×500; **b** tau-immunohistochemistry. ×500

The histological sections of the hippocampal formation showed advanced pathological changes with numerous SPs, NFTs and neuropil threads in all three cases. In general the involvement of the CA1 subregion was the most severe: it contained globular silver-positive deposits, numerous NFTs and also neurons with granulovacuolar degeneration. The tangles were typically flame shaped in the CA1 subregion, and many of them were of the ghost tangle type (Fig. 3a). The CA2 and CA3 subregions also had numerous SPs and NFTs, while the CA4 had fewer, but often very large SPs with abundant argyrophilic neurites and neurons containing rather coarse NFTs. In the outer portions of the molecular layer of the dentate fascia there was a dense row of silver-positive globular deposits (Fig. 4a). Both diffuse and compact SPs were also present inside the granular cell layer in case 3, but none of the neurons in this cell layer contained NFTs. Large numbers of SPs were seen in the subiculum, and many of the neurons of the entorhinal cortex, particularly the pre-α-neurons, contained NFTs. The amygdala contained abundant spherical amyloid deposits and frequent NFTs.

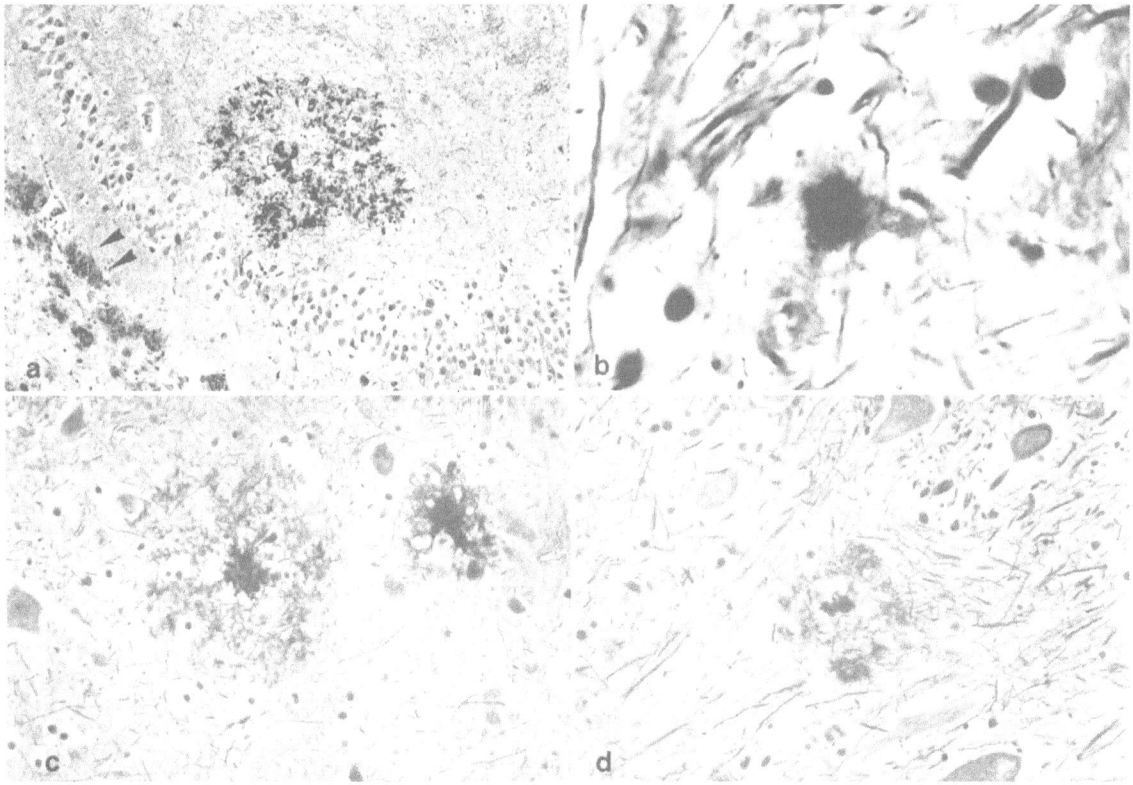

Fig. 4. Giant senile plaque rich in dystrophic neurites in the CA4. Arrowheads pointing to diffuse globular argyrophilic deposits in the dentate fascia (**a**). Small compact amyloid plaque in the globus pallidus (**b**), mature senile plaques in the dentate nucleus of the cerebellum (**c**) and spinal cord (**d**). Bielschowky silver impregnation. **a** ×120, **b** ×1,200, **c** and **d** ×300

The nucleus basalis of Meynert showed a constant involvement with depletion of nerve cells and frequent globose tangles in the remaining neurons and LBs in case 1.

In all three cases some of the large neurons of the striatum contained NFTs and in case 3 there were also SPs of the diffuse type. In this latter case rather dense amyloid deposits were present in both the external and internal segments of the pallidum whose neurons occasionally had NFTs (Fig. 4b).

The cerebellar cortex contained amyloid plaques in all cases. Diffuse argyrophilic deposits were present in the molecular layer while small and compact amyloid plaques were often seen at the border zone between the Purkinje and granular cell layers. Some of the plaques were lying deep in the granular cell layer or even in the underlying cerebellar white matter. These latter deposits often appeared on Bielschowsky preparation to have a central core surrounded by a thinner peripheral rim. In case 3 large mature SPs with abundant dystrophic neurites were found in the dentate nucleus (Fig. 4c).

In case 3 an occasional typical SP was present in the spinal cord (Fig. 4d).

In all three cases the zona compacta of the substantia nigra showed neuronal loss with free and phagocytosed neuromelanin pigment and astrocytosis. Furthermore a significant proportion of the remaining pigmented neurons possessed one or more typical LBs while others contained NFTs. The NFTs of the substantia nigra neurons usually filled the entire cell body and sometimes extended into dendrites. In case 3 occasional pigmented neurons contained both LBs and NFTs and, in addition, circumscribed Aβ positive amyloid deposits without significant numbers of surrounding dystrophic neurites were seen in the nigra (Fig. 5a). In all three cases other brainstem structures such as the periaqueductal grey matter, the anterior raphe nucleus and the pigmented brainstem nuclei including the locus coeruleus, the dorsal nucleus of the tenth cranial nerve, the paranigral and pigmented parabrachial nuclei were affected by NFT-formation.

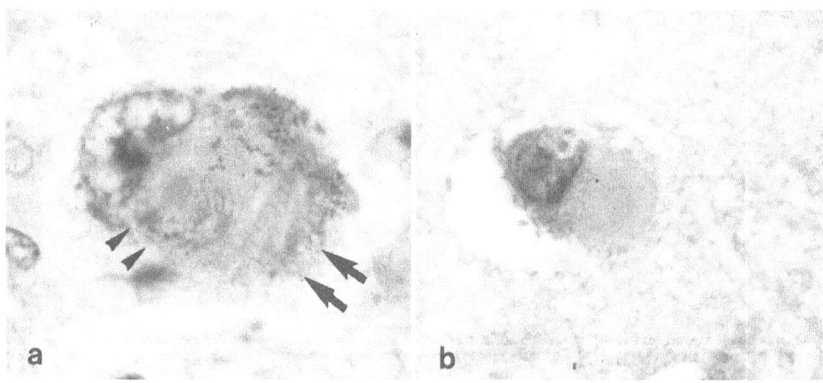

Fig. 5. Pigmented neuron in the substantia nigra containing both Lewy body (arrowheads) and neurofibrillary tangle (arrows) (**a**). Cortical neuron with Lewy body in the cingulate gyrus (**b**). **a** and **b** haematoxylin and eosin. ×1,200

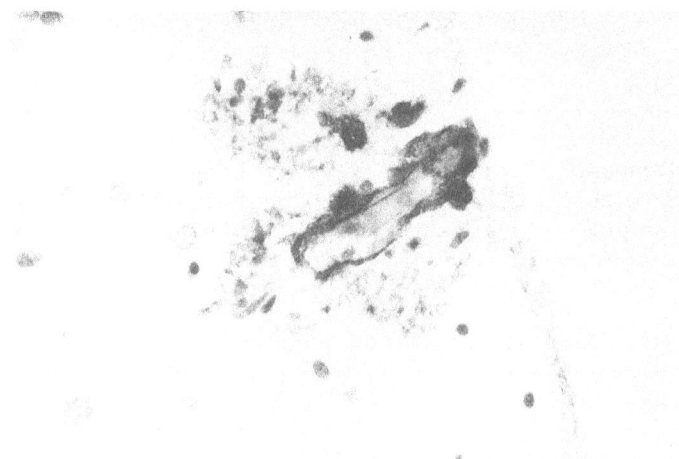

Fig. 6. Aβ-peptide positive amyloid deposition into cortical capillary. Aβ-peptide immunohistochemistry. ×500

Table 2. Neurofibrillary tangle counts (N/mm^2)

Area	Case 1	Case 2	Case 3
Frontal	14.9	29.7	50.2
Temporal	20.2	28.8	57.5
Parietal	7.7	64.4	66.3
Occipital	6.1	7.2	18.2

A proportion of the cortical neurons, especially those of the cingulate, insular and temporal cortices contained typical LBs which were numerous in cases 1 and 2, but relatively sparse in case 3 (Fig. 5b).

There was moderate to severe Aβ positive amyloid angiopathy affecting leptomeningeal as well as cerebral and cerebellar cortical small arteries and arterioles in addition to capillaries in all three cases. The vascular amyloid of the cerebral and cerebellar arterioles often spread into the surrounding neural parenchyma, and when it affected capillaries of the cerebellar cortex, the CA4 subregion of the hippocampus and some cerebral cortical areas fine perivascular spicules (drusige Entartung) were seen to be formed (Fig. 6).

The immunohistochemical study of a number of cytoskeletal components including ubiquitin, tau and phosphorylated neurofilament epitopes in case 1 showed no substantial difference compared with a sporadic case of AD (Table 1).

Morphometry

In all three cases general quantitative assessment of tangles and plaques revealed high counts in the cerebral neocortical areas examined which were well in excess of the criteria of either Khachaturian (1985) or CERAD (Mirra

Table 3. Senile plaque counts (N/mm²)

Area	Case 1	Case 2	Case 3
Frontal	57.9	13.2	60.4
Temporal	28.9	14.7	46.5
Parietal	23.4	11.7	75.0
Occipital	36.6	13.5	85.1

Table 4. Cortical Lewy body counts (N/mm²)

Area	Case 1	Average of 7 cases
Parahippocampal	3.84	1.63
Cingulate	1.46	1.53
Insular	0.76	1.15
Temporal (stg[1])	0.00	0.72
Frontal (sfg[2])	0.00	0.64
Parietal	0.76	0.46
Occipital (convex)	0.86	0.04
Calcarine	0.00	0.02

[1] *stg* superior temporal gyrus; [2] *sfg* superior frontal gyrus

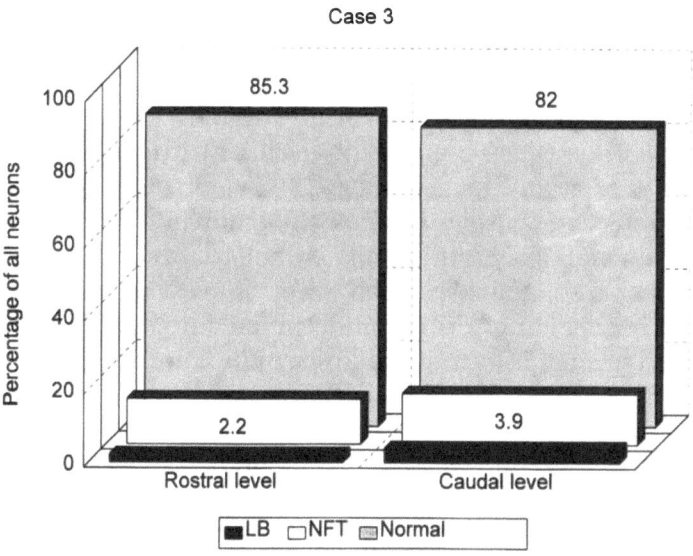

Fig. 7. Nigral neuronal cell counts and also the number of NFT and LB-bearing nerve cells were determined at two levels in case 3. The nerve cell loss was less than 5% and of those remaining about 4% contained LBs and 14% NFTs

et al., 1991) (Tables 2 and 3). In case 1 there were also high cortical LB counts (Table 4), which reached the level that is usually required for the diagnosis diffuse Lewy body disease. When compared to a normal control less than 5% loss of the pigmented neurons from the substantia nigra was noted in case 3.

Furthermore up to 14% of the remaining neurons contained NFTs and up to 4% LBs (Fig. 7).

Discussion

In all the cases presented here the neuropathological findings, including quantitative assessment, confirmed the diagnosis of AD (Khachaturian, 1985; Mirra et al., 1991; Lantos, 1992; Braak and Braak, 1994). In one case (case 3) SPs were found in "unusual" locations, such as the globus pallidus where SPs were relatively small and compact, the dentate nucleus and spinal cord, which both contained SPs rich in dystrophic neuritic processes. Furthermore some neurons of the globus pallidus contained NFTs. There was also widespread LB pathology in all three cases, primarily affecting substantia nigra and cerebral cortical neurons.

Comparative neuropathological studies have shown that familial cases are, in general, histologically indistinguishable from sporadic cases of AD, although the number of reports of systematic comparisons using quantitative methods is relatively scant. One such study, which by using semiquantitative assessment determined the severity of the pathological changes not only in ten cortical areas, but also in the amygdala, hippocampus, parahippocampal gyrus and cerebellum from 28 sporadic and 32 FAD cases, did not find any significant difference in the numbers of NFTs and SPs between familial and sporadic AD (Nochlin et al., 1993). A histological study of four brains from a large Canadian pedigree with 51 affected members came to a similar conclusion as it disclosed no "unusual" histological findings in these cases (Nee et al., 1983). In support of the previous morphological studies a detailed immunohistochemical examination of the cytoskeletal pathology, and Western blot analysis for the abnormally phosphorylated form of tau in our case 1 revealed no substantial difference between familial and sporadic AD (Lantos et al., 1992).

A number of neuropathological features have, however, been described to occur more frequently in association with FAD than with sporadic AD. Of these cerebellar amyloid plaques, which may resemble kuru plaques (Pro et al., 1980; Azzarelli et al., 1985), SPs with a greater tendency to coalesce especially in the cingulate cortex, presubiculum and striatum and, as in our case 3, compact amyloid plaques in association with amyloid angiopathy in the globus pallidus are the most frequently described (Iseki et al., 1990). The incidence of these histological abnormalities including the Aβ peptide immunoreactive and prion protein negative cerebellar plaques, is variable between different pedigrees (Martin et al., 1991). That the neuropathological findings may vary between families is further emphasised by a study which determined the density of NFTs and SPs in the hippocampal formation, including CA1–CA4, subiculum, presubiculum and dentate gyrus of eight cases from three pedigrees with FAD, all linked to DNA markers on chromosome 21 (Struble et al., 1991). Although this study was unable to find quantitative differences among the pedigrees, statistical analysis disclosed that in the CA4 subregion of the hippocampus an unusual type of SP with a marked amyloid core but without dystrophic neurites was predominantly present in two of the three pedigrees only. The presence of vacuolation or "spongy" change usually in the upper layers of the cerebral cortex has also been described in a number of pedigrees with FAD of the early onset type (Duffy et al., 1988; Martin et al., 1991).

According to the different studies the incidence of LB pathology in AD varies considerably, ranging from around 10 to above 70% (Woodard, 1962; Forno, 1982; Ditter and Mirra, 1987; Gibb et al., 1989; Kazee and Han, 1995). A recent examination of 48 cases of AD, which found the highest incidence of cortical LB pathology (71%), showed that a strong correlation exists between the number of cortical LBs and the severity of nigral pathology. This study found, however, no relationship between the incidence of cortical LBs and "plaque only" AD (Kazee and Han, 1995). Nigral and cortical LBs have been documented not only in sporadic, but also in FAD and Down's syndrome (Gibb et al., 1989; Bodhireddy et al., 1994; Raghavan et al., 1993).

The presence of LB pathology in AD highlights important nosological issues relating to the classification of neurodegenerative conditions with cortical and brainstem LBs, and addresses the question of possible overlap between AD and LB-related conditions. Brainstem LBs are classical hallmarks of Parkinson's disease, while the significance of cortical LBs in diffuse Lewy body disease was recognised just over a decade ago (Kosaka et al., 1984, 1988; Lennox et al., 1989). Furthermore it has become clear that cortical LBs are not exclusive to diffuse Lewy body disease as they have been found in nearly all cases with Parkinson's disease (Hughes et al., 1992). In the cases of diffuse Lewy body disease a widespread distribution of LBs may be associated with SPs, but NFTs are less common (Lennox, 1992). It has been suggested that, although the densities of diffuse cortical plaques may be similar in AD and diffuse Lewy body disease, "neuritic" SPs are associated with AD or with cases in which both pathologies are present (Lippa et al., 1994). A clinically and neuropathologically distinct form of LB-related conditions has also been described under the terms of "senile dementia of the Lewy body type" and "Lewy body variant of AD" (Perry et al., 1990; Hansen et al., 1989, 1990). In these cases there are many SPs and relatively few NFTs, in addition to cortical LBs which have generally a lower frequency than is observed in diffuse Lewy body disease. A possible overlap between senile dementia of the Lewy body type/Lewy body variant of AD and AD is further emphasised by the finding that the Aβ load is similar in the two conditions (Gentleman et al., 1992). Comparative examinations of the apolipoprotein E-allele distribution also revealed a relationship between AD and senile dementia of the Lewy body type/Lewy body variant of AD. These studies not only confirmed a genetic association between the ϵ4 allele and AD and showed that there is no relationship between this genetic marker and Parkinson's disease, but also demonstrated that senile dementia of the Lewy body type/Lewy body variant of AD has an ϵ4 allele distribution intermediate between AD and Parkinson's disease (Hardy et al., 1994). Furthermore apolipoprotein E ϵ4 appears to be associated with more severe neurofibrillary pathology in the senile dementia of the Lewy body type/Lewy body variant of AD cases (Hansen et al., 1994).

In the context of this rather complex neuropathological spectrum of AD and LB related conditions all our three cases should be classified as AD with brainstem and cortical LB pathology. These cases emphasise that combined phenotypic manifestations of more than one neurodegenerative condition may occur in the same cases. On the basis of our findings it is tempting to speculate that the presence of LBs in FAD and especially their occurrence in two members of the same family with the APP 717 valine to isoleucine mutation may not be pure coincidence, and to suggest that LB may be yet another manifestation of the altered cytoskeleton in AD. However the absence of LBs in other families with a similar genetic mutation also indicates that additional factors, probably both genetic and environmental, are crucial for their occurrence.

Acknowledgements

The authors wish to thank Prof. P. Luthert and Dr. T. Spargo for carrying out the morphometrical studies in cases 1 and 2. Table 1 was reproduced from the paper published by Lantos et al. [Neurosci Lett 137: 221–224 (1992)] with the permission of Elsevier Science Ireland Ltd.

References

Azzarelli B, Muller J, Ghetti B, Dyken M, Conneally PM (1985) Cerebellar plaques in familial Alzheimer's disease (Gerstmann-Straussler-Scheinker variant). Acta Neuropathol (Berl) 65: 235–246

Bird TD (1994) Familial Alzheimer's disease. Ann Neurol 36: 335–336

Bodhireddy S, Dickson DW, Mattiace L, Weidenheim KM (1994) A case of Down's syndrome with diffuse Lewy body disease and Alzheimer's disease. Neurology 44: 159–161

Braak H, Braak E (1994) Pathology of Alzheimer's disease. In: Calne DB (ed) Neurodegenerative diseases. Saunders, Philadelphia, pp 585–613

Chartier-Harlin MC, Crawford F, Houlden H, Warren A, Hughes D, Fidani L, Goate A, Rossor M, Roques P, Hardy J, Mullan M (1991) Early-onset Alzheimer's disease caused by mutations at codon 717 of the beta-amyloid precursor protein gene. Nature 353: 844–846

Corder EH, Saunders AM, Strittmatter WJ, Schmechel DE, Gaskell PC, Small GW, Roses AD, Haines JL, Pericak-Vance MA (1993) Gene dose of apolipoprotein E type 4 allele and the risk of Alzheimer's disease in late onset families. Science 261: 921–923

Ditter SM, Mirra SS (1987) Neuropathologic and clinical features of Parkinson's disease in Alzheimer's disease patients. Neurology 37: 754–760

Duffy P, Mayeux R, Kupsky W (1988) Familial Alzheimer's disease with myoclonus and "spongy change". Arch Neurol 45: 1097–1100

Forno LS (1982) Pathology of Parkinson's disease. In: Marsden CD, Fahn S (eds) Movement disorders. Butterworth Scientific, London, pp 25–40

Gentleman SM, Williams B, Royston MC, Jagoe R, Clinton J, Perry RH, Ince PG, Allsop D, Polak JM, Roberts GW (1992) Quantification of beta A4 protein deposition in the medial temporal lobe: a comparison of Alzheimer's disease and senile dementia of the Lewy body type. Neurosci Lett 142: 9–12

Gibb WR, Mountjoy CQ, Mann DMA, Lees AJ (1989) A pathological study of the association between Lewy body disease and Alzheimer's disease. J Neurol Neurosurg Psychiatry 52: 701–708

Goate A, Chartier-Harlin MCM, Brown J, Crawford F, Fidani L, Giuffra L, Haynes A, Irving N, James L, Mant R, Newton P, Rooke K, Roques P, Talbot C, Pericak-Vance M, Roses A, Williamson R, Rossor M, Owen M, Hardy J (1991) Segregation of a missense mutation in the amyloid precursor protein gene with familial Alzheimer's disease. Nature 349: 704–706

Hansen LA, Masliah E, Terry RD, Mirra SS (1989) A neuropathological subset of Alzheimer's disease with concomitant Lewy body disease and spongiform change. Acta Neuropathol 78: 194–201

Hansen L, Salmon D, Galasko D, Masliah E, Katzman R, DeTeresa R, Thal L, Pay MM, Hofstetter R, Klauber M, Rice V, Butters N, Alford M (1990) The Lewy body variant of Alzheimer's disease: a clinical and pathological entity. Neurology 40: 1–8

Hansen LA, Galasko D, Samuel W, Xia Y, Chen X, Saitoh T (1994) Apolipoprotein-E epsilon-4 is associated with increased neurofibrillary pathology in the Lewy body variant of Alzheimer's disease. Neurosci Lett 182: 63–65

Hardy J, Crook R, Prihar G, Roberts G, Raghavan R, Perry R (1994) Senile dementia of the Lewy body type has an apolipoprotein E epsilon 4 allele frequency intermediate between controls and Alzheimer's disease. Neurosci Lett 182: 1–2

Hughes AJ, Daniel SE, Kilford L, Lees AJ (1992) Accuracy of clinical diagnosis of idiopathic Parkinson's disease: a clinico-pathological study of 100 cases. J Neurol Neurosurg Psychiatry 55: 181–184

Iseki E, Matsushita M, Kosaka K, Suzuki K, Amano N, Saito A (1990) Morphological characteristics of senile plaques in familial Alzheimer's disease. Acta Neuropathol (Berl) 80: 227–232

Kazee AM, Han LY (1995) Cortical Lewy bodies in Alzheimer's disease. Arch Pathol Lab Med 119: 448–453

Khachaturian ZS (1985) Diagnosis of Alzheimer's disease. Arch Neurol 42: 1097–1105

Kosaka K, Yoshimura M, Ikeda K, Budka H (1984) Diffuse type of Lewy body disease: progressive dementia with abundant cortical Lewy bodies and senile changes of varying degree — a new disease? Clin Neuropathol 3: 185–192

Kosaka K, Tsuchiya K, Yoshimura M (1988) Lewy body disease with and without dementia: a clinicopathological study of 35 cases. Clin Neuropathol 7: 299–305

Lantos PL (1992) Neuropathology of unusual dementias: an overview. In: Rossor MN (ed) Baillière's clinical neurology, vol 1. Unusual dementias. Baillère Tindall, London, pp 485–516

Lantos PL, Luthert PJ, Hanger D, Anderton BH, Mullan M, Rossor M (1992) Familial Alzheimer's disease with the amyloid precursor protein position 717 mutation and sporadic Alzheimer's disease have the same cytoskeletal pathology. Neurosci Lett 137: 221–224

Lantos PL, Ovenstone IM, Johnson J, Clelland CA, Roques P, Rossor MN (1994) Lewy bodies in the brain of two members of a family with the 717 (Val to Ile) mutation of the amyloid precursor protein gene. Neurosci Lett 172: 77–79

Lennox G (1992) Lewy body dementia. In: Rossor MN (ed) Baillière's clinical neurology, vol 1. Unusual dementias. Baillère Tindall, London, pp 653–676

Lennox G, Lowe J, Landon M, Byrne EJ, Mayer RJ, Godwin-Austen RB (1989) Diffuse Lewy body disease: correlative neuropathology using anti-ubiquitin immunocytochemistry. J Neurol Neurosurg Psychiatry 52: 1236–1247

Levy-Lahad E, Wijsman E, Nemens E, Anderson L, Goddard KAB, Weber JL, Bird TD, Schellenberg GD (1995a) A familial Alzheimer's disease locus on chromosome 1. Science 269: 970–973

Levy-Lahad E, Wasco W, Poorkaj P, Romano DM, Oshima J, Pettingell WH, Yu C, Jondro PD, Schmidt SD, Wang K, Crowley AC, Fu Y-H, Guenette SY, Galas D, Nemens E, Wijsman EM, Bird TD, Schellenberg GD, Tanzi RE (1995b) Candidate gene for the chromosome 1 familial Alzheimer's disease locus. Science 269: 973–977

Lippa CF, Smith TW, Swearer JM (1994) Alzheimer's disease and Lewy body disease: a comparative clinicopathological study. Ann Neurol 35: 81–88

Martin JJ, Gheuens J, Bruyland M, Cras P, Vandenberghe A, Masters CL, Beyreuther K, Dom R, Ceuterick C, Lübke U, Van Heuverswijn H, De Winter G, Van Broeckhoven C (1991) Early-onset Alzheimer's disease in 2 large Belgian families. Neurology 41: 62–68

McLaughlin JE, Sankey EA, Revesz T (1993) Familial Alzheimer's disease with Lewy bodies. Neuropathol Appl Neurobiol 19: 204

Mirra SS, Heyman A, McKeel D, Sumi SM, Crain BJ, Brownlee LM, Vogel FS, Hughes JP, Van Belle G, Berg L (1991) The consortium to establish a registry for Alzheimer's disease (CERAD), part II. Standardization of the neuropathologic assessment of Alzheimer's disease. Neurology 41: 479–486

Mullan M, Crawford F (1993) Genetic and molecular advances in Alzheimer's disease. Trends Neurosci 16: 398–403

Mullan M, Crawford F, Axelman K, Houlden H, Lilius L, Winblad B, Lannfelt L (1992) A pathogenic mutation for probable Alzheimer's disease in the APP gene at the N-terminus of beta-amyloid. Nat Genet 1: 345–347

Murrell J, Farlow M, Ghetti B, Benson MD (1991) A mutation in the amyloid precursor protein associated with hereditary Alzheimer's disease. Science 254: 97–99

Nee LE, Polinsky RJ, Eldridge R, Weingartner H, Smallberg S, Ebert M (1983) A family with histologically confirmed Alzheimer's disease. Arch Neurol 40: 203–208

Nochlin D, van-Belle G, Bird TD, Sumi SM (1993) Comparison of the severity of neuropathologic changes in familial and sporadic Alzheimer's disease. Alzheimer Dis Assoc Disord 7: 212–222

Perry RH, Irving D, Blessed G, Fairbairn A, Perry EK (1990) Senile dementia of Lewy body type. A clinically and neuropathologically distinct form of Lewy body dementia in the elderly. J Neurol Sci 95: 119–139

Pro JD, Smith CH, Sumi SM (1980) Presenile Alzheimer disease: amyloid plaques in the cerebellum. Neurology 30: 820–825

Raghavan R, Khin-Nu C, Brown A, Irving D, Ince PG, Day K, Tyrer SP, Perry RH (1993) Detection of Lewy bodies in Trisomy 21 (Down's syndrome). Can J Neurol Sci 20: 48–51

Rossor M (1993a) Alzheimer's disease. BMJ 307: 779–782

Rossor M (1993b) Molecular pathology of Alzheimer's disease. J Neurol Neurosurg Psychiatry 56: 583–586

Sherrington R, Rogaev EI, Liang Y, Rogaeva EA, Levesque G, Ikeda M, Chi H, Lin C, Li G, Holman K, et al (1995) Cloning of a gene bearing missense mutations in early-onset familial Alzheimer's disease. Nature 375: 754–760

Struble RG, Polinsky RJ, Hedreen JC, Nee LE, Frommelt P, Feldman RG, Price DL (1991) Hippocampal lesions in dominantly inherited Alzheimer's disease. J Neuropathol Exp Neurol 50: 82–94

Woodard JS (1962) Concentric hyaline inclusion body formation in mental disease analysis of 27 cases. J Neuropathol Exp Neurol 21: 442–449

Authors' address: Dr. T. Revesz, Department of Neuropathology, Institute of Neurology, Queen Square, London WC1N 3BG, United Kingdom

Amyloid β-peptide and its relationship with dementia in Lewy body disease

K. Jendroska[1,2], **M. Kashiwagi**[1], **J. Sassoon**[3], and **S. E. Daniel**[4,5]

[1] Department of Neurology, Virchow-Hospital, and [2] Department of Neurology, Charité, Berlin, Federal Republic of Germany
[3] Department of Neuroimmunology, The National Hospital for Neurology and Neurosurgery, [4] The UK Parkinson's Disease Society Brain Research Centre (Brain Bank), and [5] Department of Neuropathology, Institute of Neurology, London, United Kingdom

Summary. Cerebral cortical Lewy bodies occur in a spectrum of clinical syndromes including Parkinson's disease (PD) with and without dementia, and dementing conditions clinically resembling Alzheimer's disease with few or without parkinsonian features. It is unclear whether these conditions are variants of one disease process or represent pathogenetically distinct entities. Here we compared the cortical pathology in post mortem brains of three groups representing the predominant clinical phenotypes of Lewy body disease, including 27 non-demented cases of PD, 23 demented PD cases, and 11 cases of Lewy body disease who initially presented with dementia and showed only limited features of parkinsonism during the course of their illness. In addition to neuropathology, computer-assisted histoblot analysis was used to assess cortical amyloid β-peptide deposition. There was wide overlap of the pathomorphometric features between the two groups of demented cases. It appears that substantial cortical Alzheimer-type pathology present in most demented cases contributes significantly to the development of dementia in Lewy body disease.

Introduction

Since the description of cortical Lewy bodies in cases of dementia several denominations such as Lewy body dementia (Gibb et al., 1985; Lennox, 1992), diffuse Lewy body disease (Kosaka, 1990), Lewy body variant of Alzheimer's disease (Hansen, 1990), senile dementia of Lewy body type (Perry, 1990) and others have been introduced to describe dementing conditions associated with Lewy body pathology. While some authors believe that these conditions are part of a spectrum of Lewy body disease which includes Parkinson's disease (PD) (Kosaka, 1990), others consider Lewy body dementia a distinct entity (Lennox, 1992) and try to identify specific pathogenetic factors (Benjamin et al., 1994). Many pathologists prefer the descriptive term "diffuse Lewy body disease" (DLBD) when there are numerous cortical Lewy bodies thus avoiding clinical labels as implied in PD or Lewy body dementia. The large majority of cases of DLBD described in the literature are demented and show moderate to severe Alzheimer-type changes in addition to cortical Lewy bodies (Mahler and Cummings, 1990). Here we examined the cortical pathology of 50 cases of PD, 23 of whom were demented and 27 with normal cognition. Furthermore, we compared two groups of cases who represent opposites within the spectrum of Lewy body dementia (Lennox, 1992): 23 cases of PD who developed dementia in the course of their illness and 11 cases who primarily presented with dementia, several of whom had been diagnosed clinically as probable Alzheimer's disease. Using computer-assisted densitometry on histoblots (Taraboulos et al., 1992; Jendroska et al., 1995) we analysed the distribution and density of amyloid β-peptide of senile plaques, the stability of plaques towards enzymatic degradation, and the relationship between plaques and Lewy bodies. It appears that in most demented cases of Lewy body disease co-incidental Alzheimer changes contribute significantly to dementia especially in those with a density of cortical Lewy bodies in the range of non-demented PD. However, our data show that the parameters examined are not substantially different in Lewy body dementia presenting primarily with dementia and those presenting with PD.

Patients and methods

We examined brains of 50 cases of PD and 11 cases presenting initially and primarily with dementia in whom numerous cortical Lewy bodies were found at post mortem. All cases of PD and six cases of primary dementia were obtained from the UK Parkinson's Disease Society Brain Bank; they had been followed-up by neurologists, and documentation of their mental status using the Folstein Mini-Mental State Examination and Beck Depression Index was available. A diagnosis of dementia was made following the criteria of the DSM-III-R (American Psychiatric Association, 1987). Brain tissue of five cases of primary dementia was made available through the MRC Alzheimer's Disease Brain Bank, London. Patients had been included in a prospective study on Alzheimer's disease; a clinical diagnosis of probable Alzheimer's disease was made using the NINCDS-ADRDA criteria (McKhann et al., 1984). After death, brains were obtained within 35 hours and cut mid-sagittally. One half-brain was frozen at −70°C and used for histoblot analysis; the other half-brain was fixed in 10% buffered formalin and processed for neuropathological examination. Histoblots for Aβ were carried out as described elsewhere (Taraboulos et al., 1992; Jendroska et al., 1995) using half-hemisphere frozen tissue slices of cerebrum at the level of genu corpus callosum, lateral geniculate bodies, occipital pole and of the cerebellum through the dentate nucleus. In histoblots, the progressive involvement of the cerebral cortex with Aβ immunoreactive plaques was assessed by using a three-stage classification in coronal sections at the level of the lateral geniculate body (Fig. 1): Stage I: plaques in the medial temporal lobe; Stage II: plaques in the cortex of the entire temporal lobe; Stage III: plaques in all cortical areas. To analyse the resistance of plaques to proteolytic digestion which depends on both the age and type of plaque (diffuse plaques are more susceptible to degradation than amyloid plaques), histoblots were exposed to two concentrations of protease K (50 g/ml and 400 g/ml, Boehringer) for 24 hours prior to immunostaining (Jendroska et al., 1995). Aβ staining on histoblots was quantified using a Quantimet image analyser linked to a light microscope with an

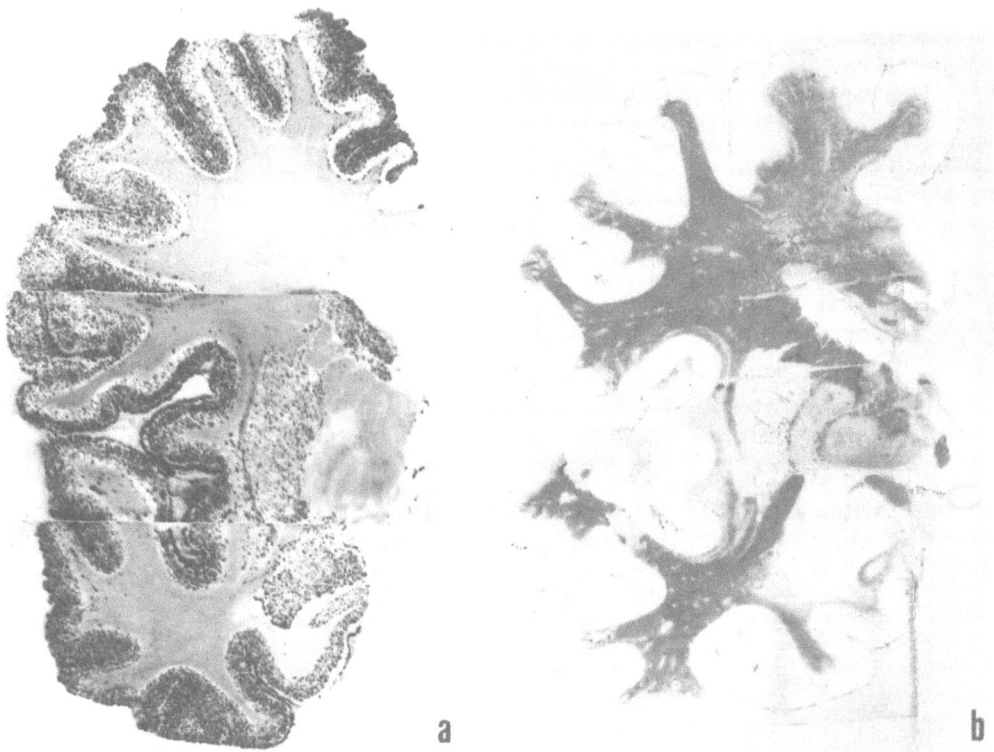

Fig. 1. Histoblots of two cases of Lewy body disease presenting initially and primarily with dementia. **a** Numerous plaques are present in the cerebral cortex, resembling Alzheimer's disease. **b** Represents a case without plaques. (Note: the discoloration of white matter in histoblots is non-specific)

adjustable objective lens. The system was calibrated before each set of measurements. Measurements were performed using a frame area of $1.2 \times 10^7 \mu m^2$, with $6 \mu m$ per pixel. Plaques appeared as dark regions on a pale background and were reproduced as a digitised image by the analysis system. An adjustable black level was used to define and quantify the Aβ deposits. Once the black level had been adjusted to encompass all Aβ immunoreactivity within a frame, the number of plaques and the total area covered with Aβ per frame was calculated. Aβ deposition was quantified by taking measurements within three random frames. The frames were placed within an area no greater than $2 cm^2$. For histology, blocks of formalin-fixed tissue were taken from multiple regions of cerebrum, cerebellar hemisphere and vermis, and brain stem. After processing to paraffin wax, sections were stained with haematoxylin and eosin, luxol fast blue-cresyl violet, and a modified Bielschowsky silver impregnation. Immunocytochemistry for Aβ (R 8271, 1:1,000, pre-treatment of sections with formic acid for 10 min), glial fibrillary acidic protein (GFAP, Dako, 1:400), Tau (Sigma, 1:1,000) and ubiquitin (Dako, 1:400) were performed using a biotin-streptavidin system. For counting of Lewy bodies regions to be scanned were demarcated by drawing around the cortical ribbon on slide coverslips over a light box. The area (in mm^2) for each region was measured using a Colormorph image analyser (Perceptive Instruments). Cortical ubiquitin-immunopositive Lewy bodies occurred predominantly in layers 5 and 6 of the cortex. They were distinguished from tangle-type inclusions which were also reactive with tau antibodies. Each area was examined at a magnification of ×100 and ×200 and density of Lewy bodies was calculated.

Results

Of 50 cases presenting with PD, 23 later suffered with dementia; in 9/23 there was severe cortical plaque pathology (Stage III) which in PD was always associated with dementia although only in 3/9 cases pathology was sufficient to warrant a diagnosis of co-incidental AD. Two PD cases with dementia and numerous cortical Lewy bodies showed neither plaques nor any other histopathological lesions to account for dementia (Table 1). Cortical Lewy bodies were present in all PD cases. Lewy body densities up to $1/mm^2$ in cingulate or parahippocampal gyrus were found in non-demented cases. Although some cases with Alzheimer-type pathology had high Lewy body counts there was no statistical relationship between the amount of $A\beta$ and the density of Lewy bodies in both cingulate and parahippocampal gyrus. The latter findings are described in more detail elsewhere (Jendroska et al., 1994). Eleven further cases presenting primarily with dementia associated with cortical Lewy bodies were examined; in addition to dementia, early symptoms in these cases included depression in four and visual or auditory hallucinations in three. Parkinsonian features were usually mild; in four cases tremor was

Table 1. Amyloid β-peptide-, plaque- and densities of cortical Lewy bodies in Lewy body dementia presenting a) primarily with dementia and b) with Parkinson's disease

Case	Age	$A\beta_{Stage}$	$A\beta_\%$*	$A\beta_{\%PK50}$	$A\beta_{\%PK400}$	Pl_d**	Pl_{dPK50}	Pl_{dPK400}	LB_{cing}***	LB_{phg}
a) Cases who presented primarily with dementia										
673	65	III	1	0.3	0	7	2	0	1	1.1
524	67	III	na	na	na	na	na	na	0.9	0.6
253	71	III	5	3	2	25	18	9	5.6	3.6
006	71	III	9	4	1.2	21	13	8	0.3	0.6
196	72	III	13	5	2	40	4	9	2.8	0.7
027	76	III	15	6	1.1	63	38	10	0.4	0.4
31/90	77	III	na	2	0.8	na	14	4	2.4	1.1
087	78	III	15	4	0.7	38	21	7	2.7	1.4
128	58	0	0	0	0	0	0	0	0.2	0.4
565	70	0	0	0	0	0	0	0	0.3	0.1
163	79	0	0	0	0	0	0	0	0.8	1
b) Cases who developed dementia in the course of Parkinson's disease										
57/90	70	III(AD)	10	4	1	29	17	9	2	0.6
49/91	72	III(AD)	20	13	0.6	50	44	8	2.1	1.2
49/90	74	III	4	2	0.3	16	19	3	1.8	1.5
15/91	75	III	na	9	0.2	na	33	3	0.3	0.1
42/89	78	III	>50	13	0.2	>50	25	3	2.6	1.4
59/90	79	III	6	7	0.2	23	30	2	1.2	1
38/92	84	III	7	7	0	24	32	0	0.2	1.3
21/89	87	III(AD)	>50	18	0.3	>50	41	2	0.1	0.1
29/92	62	0	0	0	0	0	0	0	1	0.7
18/89	74	0	0	0	0	0	0	0	1.9	1.7

* area/100 of $A\beta$-immunopositivity, before and after digestion with 50 and 400 g/ml proteinase K, average of measurements of frontal and temporal neocortex and operculum; ** $A\beta$-immunoreactive plaques/mm^2; *** Lewy bodies/mm^2; *AD* co-incidental Alzheimer's disease; *na* not available

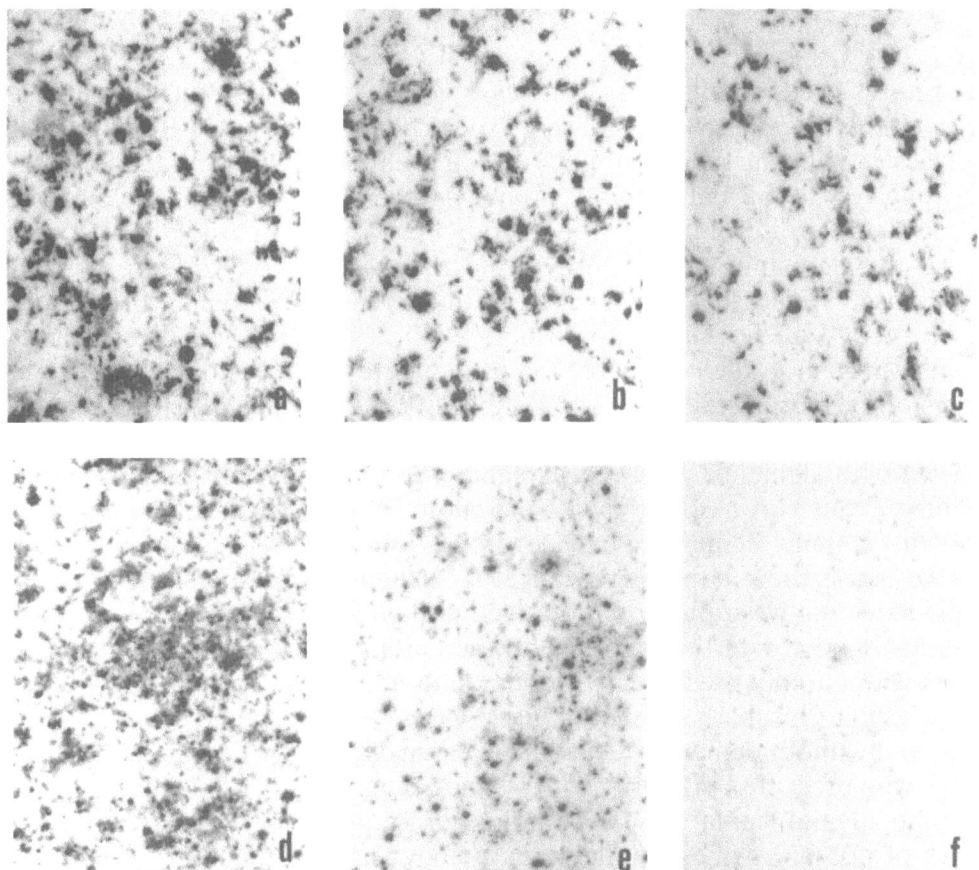

Fig. 2. Relative protease K sensitivity of predominantly neuritic (**a–c**) and predominantly diffuse plaques (**d–f**). a,d: no protease treatment; b,e: exposure to 50 mg/ml proteinase K; c,f: exposure to 400 mg/ml proteinase K. (histoblots, mag. ×500)

absent, whereas seven developed rigidity, tremor and hypokinesia. One case presented initially with auditory hallucinations and supranuclear ophthalmoplegia. Histoblot-immunostaining in these 11 cases showed Aβ plaques throughout the cerebral cortex (Stage III, Fig. 1a) in eight; in several cases the density of both Aβ immunostaining and protease-resistant amyloid plaques was comparable to cases of Alzheimer's disease (Fig. 2). Aβ densities before and after partial enzymatic digestion overlapped in cases of primary dementia and cases who developed dementia in the course of PD. Interestingly, in three cases of primary dementia aged 58, 70 and 79 there were no plaques (Fig. 1b). In one primary dementia case, Lewy body densities of 5.6/mm^2 in the cingulate and 3.6/mm^2 in the parahippocampal gyrus were exceptionally high; in the remaining primary dementia cases Lewy bodies in the cingulate gyrus ranged from 0.2 to 2.8/mm^2 with a mean of 1.2/mm^2 thus resembling demented cases of "classical" PD where the range was 0.1 to 2.6/mm^2 and the mean 1.3/mm^2. In the parahippocampal gyrus Lewy body counts in primary dementia cases were between 0.4 and 1.4/mm^2 with a mean of 0.7/mm^2 compared to demented cases of PD where the density was between 0.1 and 1.7/mm^2, the mean 1.0/mm^2.

Discussion

It is unresolved whether cases of Lewy body disease with different clinical presentations are parts of one disease spectrum or represent different disease entities in spite of their histopathological similarity. This matter is further complicated by a "babylonian" nomenclature used to describe conditions associated with Lewy bodies, relying on clinical and/or pathological criteria. Recently, Benjamin et al. (1994) found that "Lewy body disease" but not "Parkinson's disease" is associated with increased frequency of the apolipoprotein 4 allele. Three month later Arai et al. (1994) reported that the apolipoprotein 4 allele is increased in demented but not in non-demented cases of Parkinson's disease. The latter authors conclude that this increased frequency in what Lennox (1992) and others would include under the term of "Lewy body dementia" may be explained by the co-existence of Alzheimer changes (which are associated with increased frequency of the apolipoprotein 4 allele) in many dementing cases of Lewy body disease.

To assess the relative importance of Alzheimer-type pathology in Lewy body dementia we applied quantitative methodologies to cases of Lewy body disease, divided into three groups on the basis of their clinical presentation: cases who initially presented with dementia and later developed sparse features of PD ("primary dementia"), and cases who presented as classical PD with and without dementia. Comparing cases of primary dementia and those cases who presented primarily with PD and later developed dementia (Table 1) fundamental similarities of pathological parameters were found: In 8/11 cases of primary dementia plaque-pathology is severe (stage III) as in 9/23 demented PD cases. In 3/11 primary dementia cases there are no plaques as in 11/23 demented PD cases examined. Although 8/11 cases of primary dementia and 9/50 cases of PD show severe Alzheimer changes this co-existence appears co-incidental. As reported earlier (Jendroska et al., 1994) in PD we found no increase of Alzheimer pathology when compared with controls and there was no statistical relationship between the density of cortical $A\beta$ immunoreactivity and the density of Lewy bodies. Lewy body densities were comparable in demented PD and primary dementia cases. In PD, Lewy body densities of up to $1.0/mm^2$ in cingulate gyrus and/or parahippocampal gyrus are found in non-demented individuals; by analogy to PD, in 4/11 primary dementia cases Lewy body densities were in the "dementia range" ($>2/mm^2$), three cases had "borderline" Lewy body counts (approximately $1/mm^2$) but four cases were in the range of non-demented PD ($<0.5/mm^2$). Although the presence of Lewy bodies is the key histopathological feature of Lewy body dementia it is unclear how Lewy body degeneration contributes to dementia. Since in some cases of primary dementia the number of Lewy bodies found at post mortem was below that in non-demented PD cases, dementia may not be explained on the basis of Lewy body counts alone. However, our knowledge of the nature of the Lewy body in vivo is scant and the importance of the Lewy body density at post mortem should not be overrated: it is unknown how long individual Lewy bodies "survive" in the tissue. Furthermore, in analogy to the substantia nigra in PD, Lewy bodies may be confined to subpopulations of nerve cells which may get depleted thus leading to fewer Lewy bodies when the process "burns out".

In this study as well as in other series (Mahler and Cummings, 1990) it appears that Alzheimer-type changes are present in the majority of demented cases of Lewy body disease, but the relevance of this feature must be questioned in those cases in whom it is insufficient for a diagnosis of Alzheimer's disease. However, our earlier study indicated that in PD widespread Aβ deposition throughout the entire cerebral cortex is always associated with dementia even when histopathological criteria for Alzheimer's disease are not fulfilled (Jendroska et al., 1994). Hansen and Galasko (1992) suggested that the predominance of diffuse rather than neuritic plaques and the low numbers of neurofibrillary tangles in most reported cases of DLBD represent an early stage of Alzheimer's disease which may suffice for dementia when there is concomitant Lewy body degeneration. Accordingly, experimental evidence suggests that the presence of Aβ in the cerebral cortex depresses cerebral function. In vitro experiments have demonstrated that Aβ is toxic to neurons (Yankner et al., 1990) and effects apoptosis (Loo et al., 1993). In vivo, injection of specific Aβ fragments into rodent brains induces amnesia (Flood et al., 1994). Finally, as shown by Lassmann et al. (1995) in human brain there is increased nuclear fragmentation in the vicinity of senile plaques.

In summary, in Lewy body disease the majority of cases presenting initially and primarily with a dementing syndrome and approximately 40% of demented cases of "classical" PD show severe cortical Aβ deposition. Although in most cases the histopathological criteria for Alzheimer's disease are not fulfilled it appears that in the presence of Lewy-body-degeneration an earlier stage of the Alzheimer process suffices for dementia. Quantification of histological markers shows no fundamental difference between groups of Lewy body dementia with different clinical presentations. Although in clinical settings it may appear convenient to separate Lewy body disease into subgroups there is no convincing evidence for fundamental differences in the pathogenesis of these syndromes.

Acknowledgements

We wish to thank Prof. P. Lantos and Dr. N. Cairns of the MRC Alzheimer's Disease Brain Bank for contributing brain tissue of five cases. Furthermore, we thank Mrs. L. Elliott and Ms. H. Sangha for technical assistance and Ms. R. Nani for secretarial assistance.

Dr. Daniel is supported by a grant from the UK Parkinson's Disease Society.

References

American Psychiatric Association (1987) Diagnostic and statistical manual of mental disorders, 3rd ed, rev. American Psychiatric Association, Washington DC

Arai H, Muramatsu T, Higuchi S, Sasaki H, Trojanowski JQ (1994) Apolipoprotein E gene in Parkinson's disease with and without dementia. Lancet 344: 889

Benjamin R, Leake A, Edwardson JA, McKeith IG, Ince PG, Perry RH, Morris CM (1994) Apolipoprotein E genes in Lewy body and Parkinson's disease. Lancet 343: 1565

Flood JF, Roberts E, Sherman MA, Kaplan BE, Morley JE (1994) Topography of a binding site for small amnestic peptides deduced from structure-activity studies: relation to amnestic effect of amyloid β protein. Proc Natl Acad Sci USA 91: 380–384

Gibb WRG, Esiri MM, Lees AJ (1985) Clinical and pathological features of diffuse Lewy body disease (Lewy body dementia). Brain 110: 1131–1153

Hansen LA, Galasko D (1992) Lewy body disease. Curr Opin Neuro Neurosurg 5: 889–894

Hansen LA, Salmon D, Galasko D, Masliah E, Katzman R, Deteresa R, Thal L, Pay MM, Hofstetter R, Klauber M, Rice V, Butters N, Alford M (1990) The Lewy body variant of Alzheimer's disease: a clinical and pathological entity. Neurology 40: 1–8

Jendroska K, Lees AJ, Poewe W, Daniel SE (1994) Amyloid β-peptide and its relationship with dementia in Parkinson's disease. In: Korczyn AD (ed) Dementia in Parkinson's disease. Monduzzi Editore, Bologna, pp 71–78

Jendroska K , Poewe W, Daniel SE, Pluess J, Iwersen-Schmidt H, Paulsen J, Barthel S, Schelosky L, Cervós-Navarro J, DeArmond SJ (1995) Ischemic stress induces deposition of amyloid β-immunoreactivity in human brain. Acta Neuropathol 90: 461–466

Kosaka K (1990) Diffuse Lewy body disease in Japan. J Neurol 237: 197–204

Lassmann H, Bancher C, Breitschopf H, Wegiel J, Bobinski M, Jellinger K, Wisniewski HM (1995) Cell death in Alzheimer's disease evaluated by DNA fragmentation in situ. Acta Neuropathol 89: 35–41

Lennox G (1992) Lewy body dementia. Ballieres Clin Neurol 1: 653–676

Loo DT, Copani A, Pike CJ, Whittemore ER, Walencewicz AJ, Cotman CW (1993) Apoptosis is induced by β-amyloid in cultured central nervous system neurons. Proc Natl Acad Sci USA 90: 7951–7955

Mahler ME, Cummings JL (1990) Alzheimer disease and the dementia of Parkinson disease: comparative investigations. Alzheimer Dis Assoc Disord 4: 133–149

McKhann G, Grachman D, Folstein M, Katzman R, Price D, Stadlan EM (1984) Clinical diagnosis of Alzheimer's disease: report of the NINCDS-ADRDA Work Group under the auspices of Department of Health and Human Services Task Force on Alzheimer's disease. Neurology 34: 939–44

Perry RH, Irving D, Blessed G, Fairbairn A, Perry EK (1990) Senile dementia of Lewy body type: a clinical and neuropathologically distinct form of Lewy body dementia in the elderly. J Neurol Sci 95: 119–139

Taraboulos A, Jendroska K, Serban D, Yang SL, DeArmond SJ, Prusiner SB (1992) Regional mapping of prion proteins in brain. Proc Natl Acad Sci USA 89: 7620–7624

Yankner BA, Duffy LK, Kirschner DA (1990) Neurotrophic and neurotoxic effects of amyloid β protein: reversal by tachykinin neuropeptides. Science 250: 279–282

Authors' address: Dr. K. Jendroska, Department of Neurology, Charité, Schumannstrasse 20/21, D-10117 Berlin, Federal Republic of Germany

Neurochemical abnormalities in Alzheimer's disease and Parkinson's disease — a comparative review

W. Gsell, I. Strein, U. Krause, and **P. Riederer**

Department of Psychiatry, Clinical Neurochemistry, University of Würzburg, Würzburg, Federal Republic of Germany

Summary. We report a meta-analysis of the brain neurochemical abnormalities in Alzheimer's (AD) and Parkinson's disease (PD). Evidence for oxidative stress, and disorders of energy metabolism and excitatory amino acids is presented for both disorders. However, limited data and conflicting findings preclude any definitive statement relating to differences and overlap between the two conditions.

Introduction

The most prominent neuropathological finding in Parkinson's disease is a decrease in the number of the melanin-containing dopaminergic neurons in substantia nigra pars compacta. The neurochemical correlate is a marked reduction in striatal dopamine content (Ehringer and Hornykiewicz, 1960). On the other hand Alzheimer's disease is characterized by a reduction in cell number of the cholinergic basal forebrain nuclei, resulting in a cholinergic deficit in cortical regions (Perry, 1986).

We present the results of a meta-analysis of neurochemical changes in human post-mortem brains from AD and PD focusing on energy metabolism, and alterations in oxidative stress and excitatory amino acid parameters as possible etiopathogenic mechanisms. The aim of our study was to demonstrate overlaps and specific differences between the two diseases, as well as to distinguish unequivocal pathologic findings and therefore to identify either regions or neurochemical parameters which could be the subject of further post-mortem investigations.

Methods

Terminology

The term AD is used in this paper to cover presenile as well as the senile form, both severe and mild dementias, dementia of Lewy body type and familiar Alzheimer's disease. We did not distinguish between presenile versus senile, mild versus severe, or

Table 1. Single patients records from AD and PD brains for the different neurochemical systems investigated

	AD		PD		
	n	% of total AD	n	% of total PD	
Oxidative stress	318	10.5	1,330	47.4	
Energy metabolism	218	7.2	253	9.0	
Excitatory amino acids	2,489	82.3	1,222	43.6%	
Sum	3,025	100.0	2,805	100.0%	5,830

familiae versus sporadic dementias, because missing characterization in many publications would have limited the review. The term PD is restricted for idiopathic Parkinson's disease. Classification of patients was made according to DSM-III-R and/or ICD 9.

Literature used

The 59 original articles published between January 1980 and December 1993 (PD) or February 1994 (AD) cited in Medline were analyzed for post-mortem neurochemical parameters in different brain regions of Alzheimer's and Parkinson's diseases. The data pool includes 5,830 single patient records (Table 1).

Calculation

To circumvent problems caused by different units of measurement, all data were transformed to percentages of control values. Data from different authors for one parameter in a specific region were weighted according to the number of cases. Because many parameters were measured in different regions of one patient, the number of single records is given instead of the number of patients. An overall mean has been calculated using the mean from each publication with respect to the region and neurochemical parameter. This is shown in the tables, and in brackets the number of investigations and number of single records. Data are classified under "cortical regions", "basal ganglia", "limbic structures" and "other regions".

Results

Oxidative stress (Table 1)

In PD post-mortem brains, the evaluation of oxidative stress has included the measurement of ascorbic acid, free iron (either Fe(II) or Fe(III)), ferritin, glutathione (reduced (GSH) form), glutathione enzymes (transferase, peroxidases and reductases), catalase, superoxide dismutase (SOD), and malondialedehyde (MDA). In AD post-mortem brains, the number of investigated parameters was much more limited. Data were reported for glutathione (oxidized (GSSG) form) concentration and the activities of glutathione transferase and superoxide dismutases (SOD, either cytosolic or particulate).

Iron is a possible indicator of the production of reactive oxygen species. In PD there was a significant increase of free total iron, as well as of Fe(II) and especially of Fe(III) in the substantia nigra, when compared to healthy controls. But the globus pallidus also exhibited an increase, although the changes were not as high. Alterations in other areas ranged from a slight decrease in iron content within the caudate nucleus, putamen and cerebellar cortex to deviations on either side of control values within the BA 21 region of the frontal cortex and within the hippocampus. The iron binding protein ferritin was also increased in putamen and substantia nigra. Free radical damage in PD was suggested by an increase in malondialdehyde (MDA) content in nearly all regions investigated, but especially in the substantia nigra.

Within the non-enzymatic detoxication systems, ascorbic acid and reduced glutathione (total and GSH) have been investigated in PD. With the exception of the frontal cortex and the basal nucleus of Meynert (NbM), there is a general mild decrease in ascorbic acid content in PD brains, with the most prominent decrease in the amygdaloid body. Total glutathione content is decreased in all areas investigated, but changes are milder than for ascorbic acid. However, the findings differ with respect to reduced glutathione. Within the occipital and cerebellar cortices, an increase in reduced glutathione content has been reported, while in frontal cortex and substantia nigra a decrease was found.

Findings for enzymatic detoxication systems in PD are not as consistent as with the non-enzymatic ones. The most prominent finding is that for the cytosolic glutathione peroxidase (GSH-Px) enzyme, whose activity generally increases to around 300% (NbM) and up to 1300% (thalamus) of control values. Very mild increases were seen in particulate or mitochondrial GSH-Px activity, while in studies, where cytosolic and mitochondrial activity were not distinguished GSH-Px activity decreases. Glutathione reductase activity showed slight deviations to both sides of control values. The most prominent finding is an increase in activity of the cytosolic isoenzyme in NbM (157% of control value). Similar changes were reported for catalase activity, where prominent findings included an increase to 129% in substantia nigra for the cytosolic enzyme, and an increase to 190% for the particulate activity. For superoxide dismutase an increase of activity is seen in many regions, ranging up to 197 for cytoxolic and 124% for particulate activity.

In AD post-mortem brains, oxidized glutathione (GSSG) content was increased in all areas investigated, with highest concentrations in the hippocampal formation. The activity of glutathione transferase was increased, with the most significant changes in frontal cortex. Superoxide dismutase activity (SOD) shows deviations to both sides of control values, but changes were mild and not significant. The mitochondrial (particulate) isoform exhibited the strongest change in the hippocampus (115% of control values).

Table 2a. Oxidative stress parameters in AD and PD

Parameter Region	Ascorbic acid AD	PD	Free Iron AD	PD	Iron (II) AD	PD	Iron (III) AD	PD	Ferritin AD	PD	Glutathione, total AD	PD	GSH AD	PD	GSSG AD	PD	Glutathione transferase AD	PD
Cortex																		
frontal	—	101 (1,6)	—	95 (1,9)	—	—	—	—	—	—	—	86 (2,21)	—	82 (1,6)	112 (1,8)	—	128 (1,18)	110 (1,17)
BA 21	—	—	—	—	—	107 (1,6)	—	93 (1,6)	—	—	—	—	—	—	—	—	—	—
temporal	—	—	—	—	—	—	—	—	—	—	—	—	—	—	—	—	—	—
occipital	—	—	—	—	—	—	—	—	—	—	—	86 (2,15)	—	109 (1,3)	121 (1,8)	—	—	—
g. cinguli	—	—	—	89 (2,15)	—	—	—	—	—	—	—	—	—	—	—	—	—	—
cerebellar	—	—									—	83 (2,11)	—	124 (1,3)	—	—	—	—
Basal gang.																		
n. caudatus	—	—	—	93 (2,16)	—	—	—	—	—	—	—	86 (2,13)	—	—	—	—	—	—
putamen	—	69 (1,6)	—	81 (1,8)	—	72 (1,8)	—	100 (1,8)	—	137 (1,5)	—	81 (1,10)	—	—	—	—	—	—
put., lat.	—	—	—	—	—	—	—	—	—	—	—	—	—	—	—	—	—	—
put., med.	—	66 (1,6)	—	120 (1,6)	—	107 (1,6)	—	126 (1,6)	—	—	—	—	—	—	—	—	—	—
g. pallidus	—	—	—	—	—	—	—	—	—	—	—	—	—	—	—	—	—	—
pall., lat.	—	—	—	—	—	—	—	—	—	—	—	—	—	—	—	—	—	—
pall., med.	—	—	—	—	—	—	—	—	—	—	—	—	—	—	—	—	—	—
thalamus	—	71 (1,5)	—	—	—	—	—	—	—	—	—	—	—	—	—	—	—	—
Limbic sys.																		
amygdala	—	25 (1,4)	—	—	—	—	—	—	—	—	—	—	—	—	—	—	—	—
hippocampus	—	—	—	104 (1,6)	—	110 (1,6)	—	87 (1,6)	—	—	—	—	—	—	137 (1,6)	—	—	—
nbm/s. inn.	—	105 (1,3)	—	—	—	—	—	—	—	—	—	—	—	—	120 (1,8)	—	111 (1,11)	101 (1,8)
other areas																		
s. nigra	—	88 (1,3)	—	140 (2,17)	—	134 (1,8)	—	263 (1,8)	—	129 (1,5)	—	71 (2,21)	—	0 (1,6)	107 (1,7)	—	105 (1,14)	108 (1,12)
c. semiovale	—	—	—	—	—	—	—	—	—	—	—	—	—	—	—	—	—	—
hypothalamus	—	—	—	—	—	—	—	—	—	—	—	—	—	—	—	—	—	—

Table 2b. Oxidative stress parameters in AD and PD

Parameter / Region	GSH-Px, unspec. AD	PD	GSH-Px, cytosolic AD	PD	GSH-Px, particulate AD	PD	GSSG-Rd, cytosolic AD	PD	GSSG-Rd, particulate AD	PD	Catalase, cytosolic AD	PD	Catalase, particulate AD	PD	SOD, cytosolic AD	PD	SOD, particulate AD	PD	MDA AD	PD
Cortex frontal	—	81 (2,24)	—	584 (1,11)	—	96 (1,11)	—	103 (1,11)	—	90 (1,11)	—	117 (1,11)	—	100 (1,11)	—	122 (1,11)	—	111 (1,11)	—	(94) (1,10)
BA 21 temporal	—	94 (2,18)	—	1,227 (1,11)	—	106 (1,11)	—	111 (1,11)	—	88 (1,11)	—	79 (1,11)	—	91 (1,11)	—	155 (1,11)	—	115 (1,11)	—	—
occipital	—	—	—	—	—	—	—	—	—	—	—	—	—	—	—	—	—	—	—	—
g. cinguli	—	—	—	—	—	—	—	—	—	—	—	—	—	—	105 (2,28)	—	102 (2,28)	—	—	—
cerebellar	—	—	—	859 (1,11)	—	137 (1,11)	—	111 (1,11)	—	115 (1,11)	—	88 (1,11)	—	109 (1,11)	—	117 (2,22)	—	117 (1,11)	—	124 (1,10)
Basal gang. n. caudatus	—	96 (2,16)	—	630 (1,11)	—	86 (1,1)	—	92 (1,11)	—	109 (1,11)	—	103 (1,11)	—	92 (1,11)	113 (2,14)	93 (1,11)	99 (2,28)	72 (1,11)	—	111 (1,10)
putamen	—	94 (2,16)	—	532 (1,11)	—	105 (1,11)	—	114 (1,11)	—	91 (1,11)	—	84 (1,11)	—	100 (1,11)	—	103 (1,11)	—	100 (1,11)	—	—
put., lat.	—	—	—	—	—	—	—	—	—	—	—	—	—	—	—	—	—	—	—	104 (1,10)
put., med.	—	—	—	—	—	—	—	—	—	—	—	—	—	—	—	—	—	—	—	105 (1,10)
g. pallidus	—	—	—	590 (1,11)	—	100 (1,11)	—	90 (1,11)	—	110 (1,11)	—	78 (1,11)	—	100 (1,11)	—	129 (1,11)	—	87 (1,11)	—	—
pall., lat.	—	80 (2,16)	—	—	—	—	—	—	—	—	—	—	—	—	—	—	—	—	—	121 (1,10)
pall., med.	—	—	—	—	—	—	—	—	—	—	—	—	—	—	—	—	—	—	—	111 (1,10)
thalamus	—	—	—	1,286 (1,11)	—	35 (1,11)	—	95 (1,11)	—	107 (1,11)	—	103 (1,11)	—	130 (1,11)	—	127 (1,11)	—	111 (1,11)	—	—
Limbic sys. amygdala	—	—	—	829 (1,11)	—	111 (1,11)	—	112 (1,11)	—	100 (1,11)	—	100 (1,11)	—	100 (1,11)	—	100 (1,11)	—	88 (1,11)	—	—
hippocampus	—	—	—	—	—	—	—	—	—	—	—	—	—	—	93 (2,28)	—	115 (2,28)	—	—	—
nbm/s. inn.	—	—	—	306 (1,11)	—	111 (1,11)	—	157 (1,11)	—	99 (1,11)	—	118 (1,11)	—	190 (1,11)	—	197 (1,11)	—	124 (1,11)	—	—
other areas s. nigra	—	81 (2,16)	—	921 (1,11)	—	110 (1,11)	—	113 (1,11)	—	114 (1,11)	—	129 (1,11)	—	109 (1,11)	—	127 (2,22)	—	113 (2,22)	—	123 (2,28)
c. semiovale	—	—	—	696 (1,11)	—	127 (1,11)	—	91 (1,11)	—	89 (1,11)	—	110 (1,11)	—	92 (1,11)	—	98 (1,11)	—	100 (1,11)	—	—
hypothalamus	—	—	—	—	—	—	—	—	—	—	—	—	—	—	100 (2,28)	—	102 (2,28)	—	—	—

Energy metabolism (Table 2)

In PD, alterations in energy metabolism were examined by measuring the activities of pyruvate dehydrogenase, citrate synthase, isocitrate dehydrogenase, fumarase, malate dehydrogenase, and the complexes I, II, I/III, II/III, III and IV of the mitochondrial electron transport chain. In AD post-mortem brain, activities of pyruvate dehydrogenase, citrate synthase, ATP citrate lyase, fumarase, and complex II and concentrations of lactate were investigated.

The activity of the only enzyme of the glycolitic pathway examined, pyruvate dehydrogenase, was decreased in AD in all areas examined, with the greatest reduction (27%) in the occipital cortex. In PD, only one area, the putamen, was investigated for pyruvate dehydrogenase activity 1. In contrast to AD, there was no alteration. Lactate concentration in AD was slightly decreased in the caudate nucleus.

A few investigations have been reported for the enzymes of the tricarboxylic acid cycle and the initial enzyme, citrate synthase, is the best examined. In AD, its activity is reduced generally, to a minimum of 81% of control values in the amygdaloid body. In PD, decreased enzyme activity as well as no change was found, with reductions reported in the lateral globus pallidus (73% of control values). In PD, the activities of both isoforms of isocitrate dehydrogenase were decreased. Very mild changes were found in succinate dehydrogenase and fumarase activities in AD, while succinate dehydrogenase activity in PD was more affected with a decline to 71% of control values in striatum. Malate dehydrogenase activity was unaltered in PD.

The complexes of the mitochondrial electron transport chain have been studied in more detail than the tricarboxylic acid cycle enzymes in PD, but investigations in AD are lacking. Complexes I/III activity were most severely reduced in the substantia nigra, but the reduction of complex I in the amygdaloid body and of complex III in frontal cortex and striatum was nearly as marked as in substantia nigra. Findings from other areas indicated that there was a regional specific reduction in complexes I and III activity in PD.

Excitatory amino acids (Table 3)

In PD, glutamate concentrations and glutamate dehydrogenase activity have been examined. In AD, the spectrum of investigated parameters includes the glutamate concentrations and related effectors (glycine, quinoline, phencylidine) to the assessment of transmitter uptake, transporter systems and characterization of different receptor subtypes.

Glutamate content in PD was examined in the putamen and showed a very slight increase to 108% of control values. On the other hand, glutamate dehydrogenase, was markedly reduced in its activity in all regions examined. The maximum decline was found in thalamus and medial globus pallidus, although the number of investigations was very low when compared to other regions.

Table 3a. Parameters of energy metabolism in AD and PD

Parameter Region	Pyruvate DH		Citrate Synthase		ATP Citrate Lyase		Isocitrate DH, NAD⁺-dep.		Isocitrate DH, NADP⁺-dep.		Succinate DH-complex II		Fumarase		Malate DH		complex I	
	AD	PD	AD	PD	AD	PD	AD	PD	AD	PD	AD	PD	AD	PD	AD	PD	AD	PD
unspecified																		
Cortex unspecified				88 (1,4)														114 (1,4)
frontal	73 (4,38)		97 (1,10)															86 (1,5)
parietal	62 (1,13)				65 (1,13)			71 (1,5)		69 (1,5)	94 (2,28)	80 (1,5)	95 (1,5)			97 (1,5)		
temporal	69 (1,11)				63 (1,11)													
occipital	17 (1,7)																	
cerebellum				96 (1,4)														90 (1,4)
Basal gang. striatum											114 (1,3)	71 (1,5)				83 (1,5)		77 (1,5) 115 (1,4)
n. caudatus	87 (1,13)		83 (1,6)	105 (1,4)	78 (1,5)									111 (1,31)				
putamen		101 (1,31)																
g. pallidus pall., lat.				73 (1,4)														82 (1,4)
pall., med.				85 (1,4)														88 (1,4)
thalamus																		
Limbic sys. amygdala			81 (1,7)		49 (1,7)													
hippocampus hippoc. cort hippoc. for.																		
other areas s. nigra				106 (2,16)														58 (1,7)
pons white matter																		

Table 3b. Parameters of energy metabolism in AD and PD

Parameter Region	complex I/III		complex III		complex II/III		complex IV		Lactate	
	AD	PD	AD	PD	AD	PD	AD	PD	AD	PD
unspecified										
Cortex										
unspecified						108 (1,4)		94 (1,4)		
frontal				67 (1,5)				86 (1,5)	99 (2,28)	
parietal										
temporal										
occipital										
cerebellum						106 (1,4)		93 (1,4)		
Basal gang.										
striatum				65 (1,5)				71 (1,5)		
n. caudatus						92 (1,4)		101 (1,4)	80 (1,13)	
putamen										
g. pallidus										
pall., lat.						74 (1,4)		87 (1,4)		
pall., med.						97 (1,4)		96 (1,4)		
thalamus										
Limbic sys.										
amygdala										
hippocampus										
hippoc. cort										
hippoc. for.										
other areas										
s. nigra		62 (1,9)				103 (2,16)		93 (1,3)		
pons										
white matter										

In AD, with the exception of the cerebellar cortex, there was a decrease in glutamate content. The limbic system seems to be more affected than the basal ganglia, while cortical regions appeared to be relatively spared. Glutamate metabolism was also altered in AD. The glutamate decarboxylase activity was reduced in cortical regions and the limbic system, while it was increased in the basal ganglia and substantia nigra. Glutamate dehydrogenase activity was decreased in frontal cortex and elevated in occipital cortex.

Glutamate uptake and transporters were altered in AD, however, reduction in the number of uptake sites was more severe than that of transporters. Cortical uptake sites and those of the hippocampus were more affected than those of the basal ganglia. Data on excitatory amino acid transporters are lacking in the basal ganglia and the limbic system.

Findings for glutamatergic receptor systems are more complex in AD. In the frontal cortex, NMDA receptor density was greatly increased, while in all

Table 4a. Parameters of excitatory amino acids in AD and PD

Parameter Region	Aspartate uptake		EAA transporter		Glutamate receptor, unspecified		Glutamate decarboxylase		Glutamate		Glycine		Glycine receptor Kd		Glycine receptor Bmax		NMDA receptor Kd		NMDA receptor Bmax		Phencyclidine Kd		Phencyclidine Bmax	
	AD	PD	AD	PD	AD	PD	AD	PD	AD	PD	AD	PD	AD	PD	AD	PD	AD	PD	AD	PD	AD	PD	AD	PD
Cortex																								
frontal	36 (1,8)		84 (6,92)				71 (3,33)		50 (1,15); 86 (11,136)		106 (1,15)						101 (3,50)		161 (2,44)					
temporal	41 (2,11)		75 (10,148)				72 (3,32)		78 (10,94)		93 (2,19)						91 (3,50)		93 (1,44)					
parietal	39 (2,11)		76 (4,54)				88 (2,20)		99								82 (2,28)		78 (1,22)					
occipital	43 (2,11)		70 (2,32)						90 (5,65)								88 (1,22)		88 (1,22)					
g. cinguli	62 (2,6)		102 (4,64)						95 (4,53)															
cerebellar					106 (1,10)				118 (3,34) (1,12)															
Basal gang.																								
n. caudatus	55 (1,8)				163 (7,70)		125 (5,33)		89 (2,15)															
putamen	64 (1,8)				125 (3,30)		139 (2,20)		86 (2,13)	108 (1,13)														
put., lat.									64 (1,9)															
put., med.									87 (1,9)															
g. pallidus	99 (1,8)																							
pall., lat.																								
pall., med.																								
thalamus					82 (2,24)		144 (2,20)		89 (5,36)															
Limbic sys.																								
amygdala					87 (2,24)				76 (1,8)															
hippocampus	34 (1,8)				92 (34,221)		76 (4,40)		78 (7,64)		111 (1,8)		109 (4,32)		68 (4,32)		75 (2,28)		94 (1,22)		91 (4,34)	62 (4,34)		
nbm/s. inn.					86 (1,6)				82 (2,26)															
other areas																								
s. nigra							161 (2,20)		87 (2,28)															
c. semiovale																								
pons							100 (2,20)																	

Table 4b. Parameters of excitatory amino acids in AD and PD

Parameter Region	Kainate receptor Kd		Kainate receptor Bmax		Quinoline		Quisqualate receptor Kd		Quisqualate receptor Bmax		Glutamate dehydro-genase	
	AD	PD	AD	PD	AD	PD	AD	PD	AD	PD	AD	PD
Cortex												66 (8,135)
frontal	102 (1,6)				77 (2,12)						62 (2,21)	58 (8,79)
temporal	108 (1,6)				143 (2,12)							50 (6,33)
parietal	71 (1,6)				117 (1,8)							46 (6,18)
occipital											116 (1,7)	37 (1,5)
g. cinguli												
cerebellar					131 (1,4)							81 (11,128)
Basal gang.												73 (7,21)
n. caudatus	86 (2,14)		78 (1,8)									68 (25,230)
putamen					141 (1,4)							82 (19,137)
put., lat.												
put., med.												
g. pallidus												70 (7,132)
pall., lat.												55 (8,17)
pall., med.												31 (5,12)
thalamus												30 (4,15)
Limbic sys.												
amygdala												39 (1,3)
hippocampus	104 (9,70)		105 (4,32)				105 (4,32)		69 (4,32)			34 (5,7)
nbm/s. inn.												
other areas												
s. nigra												47 (45,207)
c. semiovale												
pons												

other cortical areas, as well as in hippocampus, there was a very slight decrease. Changes in the of NMDA binding sites paralleled those of receptor density. For the glycine and phencyclidine binding sites, data were reported only for the hippocampus, where the density in glycine binding was decreased to 68% of control levels and phencyclidine binding to 62% of controls. Changes in kainate receptor binding was seen in the parietal cortex, where the affinity of the ligand is decreased, and in the caudate nucleus, where a reduction in both affinity and density was reported. For the quisqualate receptor, a decrease in the density to 69% of controls was reported for the hippocampus.

Discussion

Our analysis clearly indicates that there are changes in AD and PD in parameters which measure oxidative stress, in enzymes and complexes involved in energy metabolism, and in excitatory amio acids. This may indicate that all three processes hypothesized to underlie neuronal loss in neurodegenerative disorders contribute to the etiopathogenesis of both diseases. However, conclusions regarding the relative significance of the putative underlying degenerative processes, and possible similarities, overlaps and differences can at best be tentative.

When comparing AD and PD findings, it must be considered that four-fifths of all AD investigations were examined in the excitatory amino acid system, with few data for oxidative stress (10.5%) and energy metabolism parameters (7.2%) (Table 1). For PD, studies are evenly distributed between oxidative stress (47.4%) and excitatory amino acids (43.6%). Energy metabolism parameters have been less examined (9.0%). In many cases, different parameters have been examined and neurochemical overlaps exist in oxidative stress only for glutathione transferase and the superoxide dismutase isoenzymes. However, superoxide dismutase isoenzymes were investigated in different regions in AD and PD, so that a direct comparison is impossible. Measurement of neurochemical parameters for energy metabolism have been reported for pyruvate dehydrogenase, citrate synthase, and succinate dehydrogenase. However, overlapping regions were rare. In the excitatory amino acid system, only glutamate and glutamate dehydrogenase were measured in both AD and PD. However, glutamate has been thoroughly examined in many regions of AD but in only one region in PD, and glutamate dehydrogenase was measured in many regions of PD brains but only in two regions of AD brains.

From the limited data available in the literature it can be provisionally concluded that oxidative stress parameters are slightly more altered in PD than in AD. Amongst energy metabolism parameters, pyruvate dehydrogenase activity is more affected in AD than in PD, citrate synthase activity seems to be equally altered and succinate dehydrogenase activity is more affected in PD. In the excitatory aminos acid system, glutamate content was decreased in AD, while it appears to be unchanged in PD. The converse appears true for glutamate dehydrogenase, with a greater reduction in PD. However, this latter finding must be treated cautiously, due to the small number of studies.

Unequivocal findings supporting any of the three currently fashionable hypotheses for neurodegeneration were few and far between in the present analysis. For oxidative stress, there is only assessment of superoxide dismutase isoenzyme activity in AD in cingulate cortex, caudate nucleus, hippocampus and hypothalamus; in energy metabolism, only lactate concentration in the frontal cortex of AD brains. More unequivocal findings exist for the excitatory amino acid system in AD (EAA transporter in cortical areas, and for some areas with regard to glutamate and glutamate decarboxylase and glutamate dehydrogenase activity, NMDA receptor, kainate receptor and quisqualate receptor binding) and in PD (glutamate dehydrogenase activity). All other findings must be treated with caution. However, as the blank spaces in Tables 2–4 indicate, a lot of work remains to be done.

For the classical neurotransmitter, a clear rank order of degeneration exists for each disease. In AD it is: cholinergic system > serotonergic system >> noradrenergic system > dopaminergic system. In PD, the order is reversed: dopaminergic system > noradrenergic system >> serotonergic system > cholinergic system. From the current analysis we can not construct a similar rank order of degeneration which would allow the determination of the most important pathological trigger, as many investigations which would enable such a determination remain to be done. Future investigations which convert the unequivocal findings discussed above to more reliable ones, thus filling the blank spaces in our tables, may assist the identification of pathogenetic triggers, and the development of therapeutic strategies targeted at the underlying causes of both diseases.

References

Arai H, Kobayashi K, Ichimiya Y, Kosaka K, Iizuka R (1984) A preliminary study of free amino acids in the postmortem temporal cortex from Alzheimer-type dementia patients. Neurobiol Aging 5: 319–321

Bowen DM, Davison AN, Francis PT, Palmer AM, Pearce BR, et al (1985) Neurotransmitter and metabolic dysfunction in Alzheimer's dementia: relationship to histopathological features. Interdiscipl Topics Geront 19: 156–174

Butterworth J, Tennant M, Yates CM (1988) Brain enzymes in agonal state and dementia. Biochem Soc Trans 17: 208–209

Cedarbaum JM, Sheu K, Harding B, Blass J, Agid F-J, et al (1990) Deficiency of glutamate dehydrogenase in postmortem brain samples from Parkinsonian putamen. Ann Neurol 28: 111–112

Cowburn RF, Barton AJL, Hardy JA, Wester P, Winblad B (1987) Region-specific defects in glutamate and gamma-aminobutyric acid innervation in Alzheimer's disease. Biochem Soc Trans 15: 505–506

Cowburn R, Hardy J, Roberts P, Briggs R (1988) Regional distribution of pre- and postsynaptic glutamatergic function in Alzheimer's disease. Brain Res 452: 403–407

Cowburn RF, Hardy JA, Briggs RS, Roberts P (1989) Characterisation, density, and distribution of kainate receptors in normal and Alzheimer's diseased human brain. J Neurochem 52: 140–147

Cross AJ, Crow TJ, Ferrier IN, Johnson JA (1986) The selectivity of the reduction of serotonin S2 receptors in Alzheimer-type dementia. Neurobiol Aging 7: 3–7

Cross AJ, Slater P, Simpson M, Royston C, Deakin JFW, et al (1987) Sodium dependent D-^{3}H-aspartate binding in cerebral cortex in patients with Alzheimer's and Parkinson's disease. Neurosci Lett 7: 213–217

Dexter DT, Carter CJ, Wells FR, Javoy-Agid F, Agid YJ (1989) Basal lipid peroxidation in substantia nigra is increased in Parkinson's disease. J Neurochem 52: 381–389

Ehringer H, Hornykiewicz O (1960) Verteilung von Noradrenalin und Dopamin im Gehirn des Menschen und ihr Verhalten bei Erkrankungen des extrapyramidalen Systems. Wien Klin Wochenschr 72: 1236-1239

Ellison DW, Beal MF, Mazurek MF, Bird ED, Martin JB, et al (1986) A postmortem study of amino acid neurotransmitters in Alzheimer's disease. Ann Neurol 20: 616–621

Fowler CJ, Wiberg A, Oreland L, Marcusson J, Winblad B (1980) The effect of age on the activity and molecular properties of human brain monoamine oxidase. J Neural Transm 49: 1–20

Geddes JW, Chang-Chui H, Cooper SM, Lott IT, Cotman CW (1986) Density and distribution of NMDA receptors in the human hippocampus in Alzheimer's disease. Brain Res 399: 156–161

Gilbert JJ, Kish SJ, Chang LJ, Morito C, Shannak K, et al (1988) Dementia, parkinsonism, and motor neuron disease: neurochemical and neuropathological correlates. Ann Neurol 24: 688–691

Gramsbergen JBP, Mountjoy CQ, Rossor MN, Reynolds GP, Roth M, et al (1987) A correlative study on hippocampal cation shifts and amino acids and clinico-pathological data in Alzheimer's disease. Neurobiol Aging 8: 487–494

Greenamyre JT, Penney JB, Young AB, D'Amato CJ, Hicks SP (1985) Alterations in L-glutamate binding in Alzheimer's and Huntington's disease. Science 227: 1496–1499

Greenamyre JT, Penney JB, D'Amato CJ, Young AB (1987) Dementia of the Alzheimer's type: changes in hippocampal L-^{3}H-glutamate binding. J Neurochem 48: 543–551

Hardy J, Cowburn R, Barton A, Reynolds G, Lofdahl E, et al (1986) Glutamate deficits in Alzheimer's disease. J Neurol Neurosurg Psychiatry 50: 356–357

Hardy J, Cowburn R, Barton A, Reynolds G, Lofdahl E, et al (1987) Region-specific loss of glutamate innervation in Alzheimer's disease. Neurosci Lett 73: 77–80

Jansen KLR, Faull RLM, Dragunow M, Synek BL (1990) Alzheimer's disease: changes in hippocampal N-methyl-D-aspartate, quisqualate, neurotensine, adenosine, benzodiazepine, serotonin and opioid receptors — an autoradiographic study. Neuroscience 39: 613–627

Kish SJ, Morito C, Hornykiewicz O (1985) Glutathione peroxidase activity in Parkinson's disease brain. Neurosci Lett 58: 343–346

Marklund SL, Adolfsson R, Gottfries CG, Winblad B (1985) Superoxide dismutase isoenzymes in normal brains and in brains from patients with dementia of Alzheimer type. J Neurol Sci 67: 319–325

Marttila RJ, Lorentz H, Rinne UK (1988) Oxygen toxicity protecting enzymes in Parkinson's disease. J Neurol Sci 86: 321–331

McGeer EG, Singh E, McGeer PL (1987) Sodium-dependent glutamate binding in senile dementia. Neurobiol Aging 8: 219–223

McGeer EG, Singh EA, McGeer PL (1987) Gamma-glutamyltransferase: normal cortical levels in Alzheimer disease. Alzheimer Dis Assoc Disord 1: 38–42

Mizuno Y, Suzuki K, Ohta S (1990) Postmortem changes in mitochondrial respiratory enzymes in brain and preliminary observation in Parkinson's disease. J Neurol Sci 96: 49–57

Montis de G, Beaumont K, Javoy-Agid F, Agid Y, Constandinidis JJ (1982) Glycine receptors in human substantia nigra as defined by (^{3}H)-strychnine binding. J Neurochem 38: 718–724

Moroni F, Lombardi G, Robitaille Y, Etienne P (1986) Senile dementia and Alzheimer's disease: lack of changes of the cortical content of quinolinic acid. Neurobiol Aging 7: 249–253

Mouradian MM, Contreras PC, Monahan JB, Chase TN (1988) ^{3}H-MK-801 binding in Alzheimer's disease. Neurosci Lett 93: 225–230

Palmer AM, Procter AW, Stratmann GC, Bowen DM (1986) Excitatory amino acid-releasing and cholinergic neurones in Alzheimer's disease. Neurosci Lett 66: 199–204

Pearce BR, Bowen DM (1984) ^{3}H-Kainic acid binding and choline acetyltransferase activity in Alzheimer's dementia. Brain Res 310: 376–378

Pearce BR, Palmer AM, Bowen DM, Wilcock GK, Esiri MM, et al (1984) Neurotransmitter dysfunction and atrophy of the caudate nucleus in Alzheimer's disease. Neurochem Pathol 2: 221–232

Perry EK (1986) The cholinergic hypothesis ten years on. Br Med Bull 42: 63–69

Perry E, Perry R, Tomlinson BE, Blessed G, Gibson P (1980) Coenzyme A-acetylating enzymes in Alzheimer's disease: possible cholinergic "compartment" of pyruvate dehydrogenase. Neurosci Lett 18: 105–110

Perry EK, Blessed G, Tomlinson BE, Perry RH, Crow TJ, et al (1981) Neurochemical activities in human temporal lobe related to aging and Alzheimer-type changes. Neurobiol Aging 2: 251–256

Perry TL, Yong VW (1986) Idiopathic Parkinson's disease, progressive supranuclear palsy and glutathione metabolism in the substantia nigra of patients. Neurosci Lett 67: 269–274

Perry TL, Godin DV, Hansen S (1982) Parkinson's disease: a disorder due to nigral glutathione deficiency? Neurosci Lett 33: 305–310

Perry TL, Javoy-Agid F, Agid Y, Fibiger HC (1983) Striatal GABAergic neuronal activity is not reduced in Parkinson's disease. J Neurochem 40: 1120–1123

Perry TL, Yong VW, Bergeron C, Hansen S, Jones K (1987) Amino acids, glutathione, and glutathione transferase activity in the brains of patients with Alzheimer's disease. Ann Neurol 21: 331–336

Procter AW, Palmer AM, Bowen DM, Murphy E, Neary D (1987) Glutamatergic denervation in Alzheimer's disease — A cautionary note. J Neurol Neurosurg Psychiatry 50: 825

Procter AW, Lowe SL, Palmer AM, Francis PT, Esiri M, et al (1988) Topographical distribution of neurochemical changes in Alzheimer's disease. J Neurol Sci 84: 125–140

Procter AW, Palmer AM, Francis PT, Lowe SL, Neary D, et al (1988) Evidence of glutamatergic denervation and possible abnormal metabolism in Alzheimer's disease. J Neurochem 50: 790–802

Procter AW, Stirling JM, Stratmann GC, Cross AJ, Bowen DM (1989) Loss of glycine-dependent radioligand binding to the N-methyl-D-aspartate-phencyclidine receptor complex in patients with Alzheimer's disease. Neurosci Lett 101: 62–66

Procter AW, Wong EHF, Stratmann GC, Lowe SL, Bowen DM (1989) Reduced glycine stimulation of (^{3}H)-MK-801 binding in Alzheimer's disease. Neurochem 53: 698–704

Reinikainen KJ, Paljärvi L, Huuskonen M, Soininen H, Laakso M, et al (1988) A postmortem study of noradrenergic, serotonergic and GABAergic neurons in Alzheimer's disease. J Neurol Sci 84: 101–116

Riederer P, Rausch WD, Schmidt B, Kruzik P, Konradi CJ (1988) Biochemical fundamentals of Parkinson's disease. Mt Sinai J Med 55: 21–28

Riederer P, Sofic E, Rausch WD, Schmidt B, Reynolds GP (1989) Transition metals, ferritin, glutathione, and ascorbic acid in Parkinsonian brains. J Neurochem 52: 515–520

Saggu H, Cooksey J, Dexter D, Wells FR, Lees AJ (1989) A selective increase in particulate superoxide dismutase activity in Parkinsonian substantia nigra. J Neurochem 53: 692–697

Sasaki H, Muramoto O, Kanazawa I, Arai H (1986) Regional distribution of amino acid transmitters in postmortem brains of presenile and senile dementia of Alzheimer type. Ann Neurol 19: 263–269

Schapira A, Cooper M, Dexter D, Clark J, Jenner P, et al (1990) Mitochondrial complex 1 deficiency in Parkinson's disease. J Neurochem 54: 823–827

Schapira AHV, Mann VM, Cooper JM (1990) Anatomic and disease specifity of NADH CoQ reductase (complex 1) deficiency in Parkinson's disease. J Neurochem 55: 2142–2145

Sherif F, Gottfries CG, Alafuzoff I, Oreland L (1992) Brain gamma-aminobutyrate aminotransferase (GABA-T) and monoamine oxidase (MAO) in patients with Alzheimer's disease. J Neural Transm [PD-Sect] 4: 227–240

Sheu KFR, Kim YT, Blass JP, Weksler ME (1985) An immunochemical study of the pyruvate dehydrogenase deficit in Alzheimer's disease brain. Ann Neurol 17: 444–449

Simpson MDC, Royston MC, Deakin JFW, Cross AJ, Mann DMA, et al (1988) Regional changes in ^{3}H-D-aspartate and ^{3}H-TCP binding sites in Alzheimer's disease brains. Brain Res 462: 76–82

Sofic E, Riederer P, Heinsen H, Beckmann H, Reynolds GP (1988) Increased iron (3) and total iron content in post-mortem substantia nigra of Parkinsonian brain. J Neural Transm 74: 199–205

Sofic E, Halket J, Przyborowska A, Riederer P, Beckmann H, et al (1989) Brain quinolinic acid in Alzheimer's dementia. Eur Arch Psychiatry Neurol Sci 239: 177–179

Sorbi S, Bird ED, Blass JP (1983) Decreased pyruvate dehydrogenase complex activity in Huntington and Alzheimer brain. Ann Neurol 13: 72–78

Tarbit I, Perry EK, Perry RH, Blessed G, Tomlinson BE (1980) Hippocampal free amino acids in Alzheimer's disease. J Neurochem 35: 1246–1249

Uitti RJ, Rajput AH, Rozdilsky B, Bickis M, Wollin TJ (1989) Regional metal concentrations in Parkinson's disease, other chronic neurological diseases, and control brains. Can J Neurol Sci 16: 310–314

Yates CM, Butterworth J, Tennant MC, Gordon A (1990) Enzyme activities in relation to pH and lactate in postmortem brain in Alzheimer-type and other dementias. J Neurochem 55: 1624–1630

Authors' address: Dr. W. Gsell, Department of Psychiatry, Clinical Neurochemistry, Füchsleinstrasse 15, D-97080 Würzburg, Federal Republic of Germany

Apolipoprotein E gene in Parkinson's disease, Lewy body dementia and Alzheimer's disease

D. St. Clair

Department of Mental Health, University of Aberdeen, United Kingdom

Summary. Apolipoprotein E (Apo E) epsilon 4 allele is a risk factor for early and late onset Alzheimer's disease. This prompted us to examine other neurophyschiatric phenotypes. Epsilon 4 allele was significantly enriched in Lewy body dementia (N = 39) but not in Parkinson's disease (n = 50) or Schizophrenia (n = 175) compared to aged non-demented controls (n = 47) and the Scottish population (n = 400). We conclude that Lewy body disease should be regarded as a variant of Alzheimer's but not Parkinson's disease.

Introduction

Alzheimer's disease (AD) is a neurodegenerative disease characterised by β amyloid protein (βAP) deposits in senile plaques and cerebral blood vessel walls, and neurofibrillary tangles (NFTs) within neurones of the cerebral cortex and hippocampus. Lewy body dementia (LBD), also known as diffuse Lewy body disease has been reported as a common cause of dementia in the elderly (Byrne et al., 1989; Dickson et al., 1989; Hansen et al., 1990). In common with AD most cases show the neuropathological features of amyloid plaques, but unlike AD they show few or no neurofibrillary tangles. It is currently controversial whether LBD is a distinct entity or a variant form of Parkinson's disease or AD. For this reason, some prefer to use the term Lewy body variant (LBV) of AD. However, LBD does have distinguishing clinical and pathological features: fluctuating confusional states or psychoses are presenting features; the neuritic plaques lack the cytoskeletal abnormalities found in AD2, cortical neurofibrillary tangles are sparse or absent and cortical Lewy bodies (LBs) have a high level of ubiquitin c terminal hydroxylase not found in other inclusion bodies such as neurofibrillary tangles (Lowe et al., 1990). In contrast to Parkinson's disease much has recently been discovered about the molecular biology of AD. At least five susceptibility loci must be involved, four of which have been identified and characterised, amyloid precursor protein (APP) gene on chromosome 21 (Goate et al., 1991), the recently cloned presenilin I gene on chromosome 14 (Sherrington et al., 1995) and presenilin 2 gene on chromosome 1 (Levy-Lahad et al., 1995) and Apolipoprotein E (ApoE) (Strittmatter et al., 1993) gene on chromosome 19. Pathogenic mutations have been identified in family members affected by early onset autosomal dominant Alzheimer's disease in APP gene, presenilin I and presenilin 2 genes. Fortunately families harbouring such mutations are rare so that even when combined, are responsible for only a very small proportion of AD as a whole (Van Broeckhoven, 1995).

ApoE is a sialoglycoprotein involved in cholesterol transport. There are three isoforms E2, E3 and E4 that are the products of three alleles (ε2, 3 and 4) located on the long arm of chromosome 19. Astrocytes and microglia synthesise most of the ApoE found in brain. Neurones do not produce ApoE but express an ApoE binding site, low density lipoprotein receptor related protein by which it can be internalised. ApoE has now been unequivocally implicated in Alzheimer's disease a) ApoE 4 in RNA levels are increased in AD brains relative to controls (Diedrich et al., 1991) b) ApoE is associated immunochemically with plaques and neurofibrillary tangles c) ApoE has high avidity binding to A4 peptide and may promote aggregation (Nambay et al., 1991) d) e4 allele is enriched in late onset AD populations (Strittmatter et al., 1993). This latter observation has now been confirmed by dozens of groups world wide (Roses, 1995). The ApoE ε4 allele is therefore a major genetic risk factor for Alzheimer's disease. In contrast to mutations of APP and presenilin genes, ε4 allele is neither sufficient nor necessary to cause Alzheimer's disease i.e. many ε4 allele carriers do not develop AD and many non ε4 carriers do develop AD.

Methods

These exciting findings of raised ε4 allele frequency in late onset AD prompted us to examine ε4 allele frequency in other neuropsychiatric phenotypes including early onset AD, Lewy body dementia, Parkinson's disease and schizophrenia as well as neuropathologically validated aged non-demented controls. In particular, we argued that similarities or differences in allele frequencies, compared with late onset AD could help resolve the controversy regarding the status of early onset AD and LBD as separate disease entities.

Patient and normal DNA samples were as follows:

1. Early onset AD (n = 85) mean age of presentation 57 yrs (SD6) (28M, 57F). All met NINCDS criteria for definite (n = 19) or probable AD (n = 66). Other features are described elsewhere (St. Clair et al., 1995).
2. Late onset AD (n = 68) mean age of death 81 (range 69–91). All died in hospital and had clinical and neuropathological features to satisfy NINCDS criteria for a diagnosis of definite AD (McKhann et al., 1984).
3. Lewy body dementia (LBD) (n = 39) mean age of death 79 (range 66–190). All died in either psychiatric or geriatric wards. Most (n = 29) were obtained from Newcastle General Hospital. Post mortem examination revealed LBs in the substantia nigra and locus ceruleus (St. Clair et al., 1994).
4. Parkinson's disease (n = 58) mean duration of illness 17 yrs (range 5–36 yrs), mean age at death 78 (range 69–90). (39M, 19F). All were examined by Dr. Daniel at

Table 1. Frequencies of ε2, ε3 and ε4 apolipoprotein E alleles

	ε2	ε3	ε4
Early onset AD n = 85	0.06	0.53	0.41
Late onset AD n = 68	0.01	0.60	0.38
Lewy body dementia n = 39	0.04	0.61	0.35
Parkinson's disease n = 50	0.15	0.65	0.19
Schizophrenia n = 175	0.07	0.76	0.15
Non demented controls n = 47	0.03	0.84	0.13
Scottish population n = 400	0.08	0.77	0.15

Parkinson's Disease Society Brain Bank and met neuropathological criteria for idiopathic Parkinson's disease. Most have been described elsewhere (Hughes et al., 1992).

5. Schizophrenia (n = 175). All met RDC and DSMIV criteria for schizophrenia. A full description of the clinical features of the sample is in preparation.

6. Aged, non-demented controls (n = 47). The mean age at death was 78 (range 67–91) (29M, 18F). All had full neuropathological examination and showed no features of AD. Population data were also available for ApoE allele frequencies in Scotland (n = 400) mean age was 53 (SD 0.22).

ApoE alleles were determined on genomic DNA samples by polymerase chain reaction (PCR) amplification of ApoE gene sequences followed by restriction isotyping. The method used was essentially that described by Wenham et al. (1991).

Results

Table 1 gives ApoE allele frequencies for each of the groups. Highly significant differences (p < 0.001) were observed in early onset AD, late onset AD and LBD groups compared to the Parkinson's, schizophrenia and normal control groups. On the other hand, no significant differences were observed between either AD groups and LBD group. Indeed their allele distributions were remarkably similar.

These findings demonstrate that ApoE ε4 is a major risk factor for early onset and late onset AD. It is also a risk factor for LBD but not Parkinson's disease or schizophrenia. It seems likely that AD and LBD share at least a degree of similarity in aspects of basic aetiology, and this may be due to the AD neuropathological features common to both groups. On the other hand it is difficult to see where the Lewy bodies fit in. If Lewy bodies

themselves were crucial, one might expect ε4 also to be enriched in
Parkinson's disease. Results highlight the need for ApoE genotyping to be
performed on a variety of neuropsychiatric disorders, preferably with post
mortem validation, in order that its specificity and hence clinical role can
be properly evaluated. Furthermore, these observations highlight the value
of molecular biology in helping to define the proper classification of
neuropsychiatric disorders.

Acknowledgements

Dr. M. Rennie, Dr. C. Yates and Dr. R. Perry for clinical samples, Dr. S. E. Daniel and
Dr. A. Gordon for neuropathological examinations. Much of the work was conducted at
MRC Human Genetics Unit, Edinburgh and supported by The Wellcome Trust.

References

Byrne EJ, Lennox G, Lowe J, et al (1989) Diffuse Lewy body disease: clinical features in
    15 cases. J Neurol Neurosurg Psychiatry 52: 709–717
Diedrich JF, Minnigan H, Carp RI, et al (1991) Neuropathological changes in scrapie and
    Alzheimer's disease are associated with increased expression of apolipoprotein E and
    cathepsin D in astrocytes. J Virol 65: 4759–4768
Dickson DW, Crystal H, Mattiace LA, et al (1989) Diffuse Lewy body disease: light and
    electron microscopic immunohistochemistry of senile plaques. Acta Neuropathol 78:
    572–84
Goate A, Chartier HM, Mullan M, et al (1991) Segregation of a missense mutation in the
    amyloid precursor protein gene with familial Alzheimer's disease. Nature 349: 704–
    706
Hansen L, Salmon D, Galasko D, et al (1990) The Lewy body variant of Alzheimer's
    disease: a clinical and pathological entity. Neurology 40: 1–8
Hughes AJ, Daniel SE, Kilford L, et al (1992) Accuracy of clinical diagnosis of idiopathic
    Parkinson's' disease: a clinicopathological study of 100 cases. J Neurol Neurosurg
    Psychiatry 55: 181–184
Levy-Lahad L, Wijsman EM, Nemens E, et al (1995) A familial Alzheimer's disease locus
    on chromosome 1. Science, 269: 980–997
Lowe J, McDermott H, Landon M, Mayer RJ, Wilkinson K (1990) Ubiquitin carboxyl
    terminal hydroxylase (PDP9.5) is selectively present in ubiquitinated inclusion bodies
    characteristic of human neurodegenerative diseases. J Pathol 161: 153–160
McKhann G, Drachman D, Folstein M, et al (1984) Clinical diagnosis of Alzheimer's
    disease: report of the NINCDS-ADRDA work group. Neurology 34: 939–944
Namba Y, Tomonawa M, Kawasaki H, et al (1991) Apolipoprotein E immunoreactivity
    in cerebral amyloid deposits and neurofibrillary tangles in Alzheimer's disease and
    kuru plaque amyloid in Creutzfeld Jakob disease. Brain Res 541: 163–166
Roses AD (1995) Apolipoprotein E genotyping in the differential diagnosis, not predic-
    tion of Alzheimer's disease. Ann Neurol 38: 6–14
Sherrington R, Rogaev EI, Liang Y, et al (1995) Cloning of a gene bearing missense
    mutations in early onset familial Alzheimer's disease. Nature 375: 754–760
St Clair D, Norman J, Perry R, et al (1994) Apolipoprotein E epsilon 4 allele frequency
    in patients with Lewy body dementia, Alzheimer's disease and age-matched controls.
    Neurosci Lett 176: 45–46

St Clair D, Rennie M, Slorach. E, et al (1995) Apolipoprotein E epsilon 4 allele is a risk factor for familial and sporadic presenile Alzheimer's disease in both home and heterozygote carriers. J Med Gen 32: 642–644

Strittmatter WJ, Saunders AM, Schmechel D, et al (1993) Apolipoprotein E: high-avidity binding to β-amyloid and increased frequency of type 4 allele in late-onset familial Alzheimer disease. Proc Natl Acad Sci USA 90: 1977–1981

Van Broeckhoven C (1995) The presenilins. Nature Genet 11: 230–232

Wenham PR, Price WH, Blundell A (1991) Apolipoprotein E genotyping by one stage PCR. Lancet 337: 1158–1159

Author's address: Dr. D. St. Clair, Department of Mental Health, University of Aberdeen, Aberdeen AB92ZD, United Kingdom

Indices of oxidative stress in Parkinson's disease, Alzheimer's disease and dementia with Lewy bodies

A. D. Owen[1]**, A. H. V. Schapira**[2,3]**, P. Jenner**[1]**, and C. D. Marsden**[3]

[1]Neurodegenerative Disease Research Centre, Pharmacology Group, Biomedical Sciences Division, King's College, [2]Department of Clinical Neurosciences, Royal Free Hospital Medical School, and [3]University Department of Clinical Neurology, Institute of Neurology, National Hospital for Neurology and Neurosurgery, London, United Kingdom

Summary. The cause of neuronal cell death in Parkinson's disease is unknown but there is accumulating evidence suggesting that oxidative stress may be involved in this process. Current evidence shows that in the substantia nigra there is altered iron metabolism, decreased levels of reduced glutathione and an impairment of mitochondrial complex I activity. However, these changes seem to be unique to the substantia nigra and have not been found in other areas of the brain known to be altered in Parkinson's disease, such as substantia innominata. In addition they do not appear to be related to the presence of Lewy bodies, as other areas of the brain containing Lewy bodies do not show evidence of either oxidative stress or mitochondrial dysfunction. Oxidative stress has now been demonstrated in Alzheimer's disease and its presence appears to be correlated with regions of marked pathological changes.

Introduction

In Parkinson's disease (PD) there is a selective loss of dopaminergic neurones from the substantia nigra pars compacta, with characteristic eosinophilic intraneuronal Lewy bodies in remaining nerve cells. Lewy bodies are also found in other regions of the PD brain, such as the locus coeruleus, substantia innominata and cerebral cortex. Lewy bodies are also found in the cerebral cortex in dementia with Lewy bodies (DLB), (see this volume; Perry et al., 1996). Senile dementia of the Alzheimer's type (SDAT) is the most common cause of dementia in the elderly, but dementia with Lewy bodies appears to be the second commonest aetiology.

Nigral neuronal loss in PD is associated with alteration in indices of oxidative stress with inhibition of the mitochondrial respiratory chain, (see Jenner, 1994 for review). In this paper we review studies of other regions of the PD brain in which Lewy bodies are known to be present in order to test the hypothesis that oxidative stress is intimately related to their presence. Comparisons are made with indices of oxidative stress in SDAT and DLB.

Evidence for oxidative stress in Parkinson's disease

Oxidative stress in the substantia nigra in PD is suggested by the observation of increased levels of total iron (Dexter et al., 1990; Reiderer et al., 1989) loss of reduced glutathione (GSH) content (Reiderer et al., 1989; Sian et al., 1994) and reduced concentration of α-ketoglutarate dehydrogenase (α-KGDH), a key TCA cycle enzyme, as measured by immunohistochemistry (Mizuno et al., 1994). More direct evidence of nigral damage due to oxidative stress has been obtained by the findings of increased lipid peroxidation (Dexter et al., 1989) and DNA damage (Sanchezramos et al., 1994). These changes are accompanied by a significant decrease in the activity of substantia nigra complex I activity (Schapira et al., 1990a). The changes in the levels of iron and GSH, and in the activity of complex I may contribute to neuronal loss, so have been the basis of the present investigation.

Iron

In the substantia nigra of patients with PD there is an increase in total iron levels of some 35% (Dexter et al., 1987, 1989; Sofic et al., 1988; Reiderer et al., 1989). Free reactive iron can catalyse the decomposition of lipid peroxides and the formation of reactive oxygen species.

In brain, iron is mainly stored bound to ferritin which renders it less reactive. However, in PD there is controversy over the changes which occur in ferritin levels. There are reports from these laboratories of a decrease (Dexter et al., 1990) or no change in ferritin, measured using antibodies to both H and L forms of ferritin within substantia nigra (Mann et al., 1994). However, other studies have reported increased ferritin levels, although they may have been studying cases of Parkinson's-plus syndromes and not idiopathic PD.

Increases in the levels of iron are also found in substantia nigra in progressive supranuclear palsy (PSP) and multiple system atrophy (MSA) (Dexter et al., 1993), suggesting that the changes in iron levels in PD are not a cause of the disease but are a result of the neurodegenerative processes. Indeed alterations in iron levels occur in the caudate-putamen in Huntington's disease (HD) and in a variety of other neurodegenerative diseases and disease pathologies.

Glutathione

In the substantia nigra in PD a decrease in the levels of GSH occurs which would correlate with the increased formation of reactive free radicals. Initial reports of decreased GSH levels (Perry et al., 1982) were criticised because of non-physiological ratios between reduced and oxidised glutathione in control brains and the total absence of glutathione in the PD. However, other studies have confirmed a decrease in GSH levels in substantia nigra in PD (Reiderer et al., 1989; Sian et al., 1994). Sian et al. (1994) found a decrease of some 40% in the GSH levels, but no changes in oxidised glutathione (GSSG) within the substantia nigra, but not in any other regions of the PD brain. Unlike the changes seen in nigral iron levels, the reduction of nigral GSH was shown to be limited to PD, and was not shown in PSP, MSA or HD (Sian et al., 1994). This suggests that the changes in GSH are not a non-specific consequence of the neurodegeneration, but may be related to the primary pathological process. In addition there is known to be a specific increase in the activity of γ-glutamyl transpeptidase (γ-GTP), the enzyme responsible for the degradation and translocation of glutathione (Sian et al., 1994). Its elevation may indicate increased removal of GSSG from cells leading to an irreversible depletion of GSH. Alternatively γ-GTP can act to conserve precursors of GSH and its induction may be an attempt to reverse the decrease in cellular GSH content.

Mitochondrial function

Since MPTP is capable of inhibiting mitochondrial complex I, the activity of mitochondrial enzymes was examined in substantia nigra in PD brains. No changes in markers for complexes II/III or IV were found, but the activity of NADH CoQ$_1$ reductase (complex I) was reduced (Schapira et al., 1990a). This change was specific to the substantia nigra, and was not seen in MSA (Schapira et al., 1990b). To date, no defects have been found in mitochondrial DNA coding for complex I. Immunohistochemical staining has shown that the TCA cycle enzyme α-KGDH is present in decreased concentrations in the substantia nigra in PD (Mizuno et al., 1994). Since α-KGDH supplies substrate for complex II, this may affect the activity of the respiratory chain. Interestingly, α-KGDH is also inhibited by MPP$^+$.

Oxidative stress in incidental Lewy body disease

The data obtained from cases of advanced PD described above do not reveal whether alterations in indices of oxidative stress results from neuronal death in substantia nigra or whether it is a primary cause of degeneration occurring early in the disease. Studies using brain tissues from patients with incidental Lewy body disease (ILBD) have helped resolve this issue. ILBD may represent presymptomatic PD since individuals are normal but are found to exhibit some cell loss in substantia nigra accompanied by Lewy bodies at postmortem. In ILBD, the levels of iron and ferritin in substantia nigra are unchanged, again suggesting that alterations in iron metabolism are a secondary and non-specific component of neurodegeneration. Complex I activity in substantia nigra in ILBD is also unchanged, but the levels of GSH are decreased to the same extent as in advanced PD (Dexter et al., 1994). The loss of GSH is the earliest biochemical marker of nigral cell loss so far recorded. Why GSH levels are decreased is unclear. A lack of increased levels of GSSG does not fit easily with the oxidative stress concept of the disorder. Rather, decreased GSH levels may be an early indicator of impaired mitochondrial function as previously shown in hepatocytes exposed to a range of mitochondrial poisons.

Relationship between Lewy body pathology and oxidative stress in Parkinson's disease

The substantia innominata also degenerates in PD and was the area in which Lewy originally observed the cellular inclusion which now bears his name. However, iron levels in substantia innominata in PD are not different from those of age-matched controls. Levels of GSH and the activity complex I are also unchanged. So, pathological changes resulting in Lewy body formation do not *per se* involve oxidative stress. However, the level of pathology which occurs in substantia innominata in PD is less marked than in substantia nigra, so small alterations in indices of oxidative stress may not be observed. An important difference between substantia innominata and substantia nigra is the nature of the transmitters utilised. Alterations in dopamine metabolism in substantia nigra may also be key to the onset and progression of oxidative stress in substantia nigra whereas this is not a prominent feature with acetylcholine in substantia innominata. The degeneration of the noradrenalin-producing locus coeruleus in PD might provide a clue to the involvement of altered catecholamines in altered indices of oxidative stress, but it has not been possible to obtain tissues for this purpose.

Oxidative stress in dementia with Lewy bodies

The pathological characteristics of DLB are discussed in detail elsewhere in this volume and Perry et al. (1996). In cingulate cortex in DLB the levels of iron are increased. However, there is no change in GSH level or in complex I activity. These findings are interesting since they appear to separate the necessity for alterations in GSH and complex I from alterations in iron metabolism. Although, again the latter may merely be indicative of pathological abnormality. So, again there appears to be no simple association between a pathological process involving Lewy body formation and alterations in indices of oxidative stress or mitochondrial function. However, relatively few studies of DLB brain have been undertaken and many of the markers of oxidative stress and oxidative damage measured in PD have not been investigated in this disorder.

In PD patients, no evidence of alterations in iron levels, GSH levels and complex I activity are found in cingulate cortex. This would be expected as the degree of pathology in this brain area in PD is minimal.

Evidence for oxidative stress in Alzheimer's disease

The substantia innominata shows marked pathology in SDAT but accompanied by the presence of neurofibrillary plaques and tangles rather than Lewy bodies. In the substantia innominata in SDAT there was a small but non-significant loss of iron. In contrast, the level of GSH was reduced but we have not so far measured complex I activity in this brain region. It has proved difficult to interpret these changes. Neuronal loss without iron deposition is unusual and more detailed studies of iron distribution need to be undertaken. The loss of GSH may indicate the involvement of free radical species although it may simply reflect neuronal loss since a large proportion of GSH is located within neuronal cell bodies. Unless other indices of oxidative stress and damage are measured no firm conclusions can be reached. If oxidative stress does occur it will confirm the idea that the process occurs in a range of degenerative illnesses irrespective of their primary cause.

Significant increases in the levels of iron and a decrease in the level of GSH in the cingulate cortex occur in cingulate cortex in SDAT, although complex I appears normal.

There is other evidence to support alteration in indices of oxidative stress in the cortex in SDAT. Lipid peroxidation is increased in the inferior temporal cortex (Palmer and Burns, 1994) and damage to mitochondrial DNA in the parietal cortex is also increased (Mecocci et al., 1994). In addition to these markers of oxidative stress there is also evidence that mitochondrial function may be compromised. Decreases in the level of α-KGDH (Mastrogiacomo et al., 1993) and the activity of complex IV (Parker and Parks, 1994) have both been shown in SDAT.

Oxidative stress may be a factor in the degeneration which occurs in the substantia innominata and cingulate cortex of SDAT. The occurrence of oxidative stress would appear to be selective in SDAT, as in PD. Thus, studies on the temporal cortex in SDAT, an area which is relatively spared from the pathological process show no changes in iron or GSH levels. Investigation of complex I activity in this region is currently under way.

Conclusions

From the results of this study it would appear that the presence of pathology characterised by Lewy bodies and the markers of oxidative stress and mitochondrial function which have been examined are not intimately connected, either in the substantia innominata in PD or the cingulate cortex in DLB. However, evidence of oxidative stress and mitochondrial dysfunction are seen in these areas in SDAT. It is possible that the lack of evidence for oxidative stress is due to the degree of pathology not being great enough to be measured against a non-pathological background. Although this would not explain why decreases in GSH levels are seen in ILBD, where pathology is not extensive.

Acknowledgements

The support of the Medical Research Council, the Parkinson's Disease Society of the United Kingdom and the National Parkinson Foundation, Miami, U.S.A., are gratefully acknowledged. We would also like to acknowledge the help of The Parkinson's Disease Society Brain Research Centre (Brain Bank), The MRC Alzheimer's Disease Brain Bank and The Institute for the Health of the Elderly at Newcastle General Hospital, for supplying brain materials for this study.

References

Dexter DT, Wells FR, Agid F, Agid Y, Lees AJ, Jenner P, Marsden CD (1987) Increased nigral iron content in post-mortem parkinsonian brain. Lancet ii: 219–220

Dexter DT, Carter CJ, Wells FR, Javoy-Agid F, Agid Y, Lees A, Jenner P, Marsden CD (1989) Basal lipid peroxidation in substantia nigra is increased in Parkinson's disease. J Neurochem 52: 381–389

Dexter DT, Wells FR, Lees AJ, Agid F, Agid Y, Jenner P, Marsden CD (1989) Increased nigral iron content and alterations in other metal ions occurring in the brain in Parkinson's disease. J Neurochem 52: 1830–1836

Dexter DT, Carayon A, Vidaillhet M, Ruberg M, Agid F, Agid Y, Lees AJ, Wells FR, Jenner P, Marsden CD (1990) Decreased ferritin levels in brain in Parkinson's disease. J Neurochem 55: 16–20

Dexter DT, Sian J, Jenner P, Marsden CD (1993) Implications of alterations in trace element levels in brain in Parkinson's disease and other neurological disorders affecting the basal ganglia. Adv Neurol 60: 273–281

Dexter DT, Sian J, Rose S, Hindmarsh JG, Mann VM, Cooper JM, Wells FR, Daniel S, Lees AJ, Schapira AHV, Jenner P, Marsden CD (1994) Indices of oxidative stress and mitochondrial function in individuals with incidental Lewy Body disease. Ann Neurol 35: 38–44

Jenner P (1994) Oxidative damage in neurodegenerative disease. Lancet 344: 796–798

Mann VM, Cooper JM, Daniel SE, Srai K, Jenner P, Marsden CD, Schapira AHV (1994) Complex I, iron and ferritin in Parkinson's disease substantia nigra. Ann Neurol 36: 876–881

Mastrogiacomo F, Bergeron C, Kish SJ (1993) Brain α-ketoglutarate dehydrogenase complex activity in Alzheimer's disease. J Neurochem 61: 2007–2014

Mecocci P, MacGarvey U, Flint-Beal M (1994) Oxidative damage to mitochondrial DNA is increased in Alzheimer's disease. Ann Neurol 36: 747–751

Mizuno Y, Matsuda S, Yoshina H, Mori H, Hattori N (1994) An immunohistochemical study of α-ketoglutarate dehydrogenase complex in Parkinson's disease. Ann Neurol 35: 204–210

Palmer AM, Burns MA (1994) Selective increase in lipid peroxidation in the inferior temporal cortex in Alzheimer's disease. Brain Res 645: 338–342

Parker WD, Parks JK (1995) Cytochrome c oxidase in Alzheimer's disease brain: purification and characterization. Neurol 45: 482–486

Perry T, Godin DV, Hansen S (1982) Parkinson's disease: a disorder due to nigral glutathione deficiency. Neurosci Lett 33: 305–310

Perry E, McKeith I, Perry R (1996) Dementia with Lewy bodies: clinical, pathological and treatment issues. Cambridge University Press

Reiderer P, Sofic E, Rausch W-D, Schmidt B, Reynolds GP, Jellinger K, Youdim MB (1989) Transition metals, ferritin, glutathione and ascorbic acid in Parkinson's disease. J Neurochem 52: 381–389

Sanchezramos JR, Overvik E, Ames BN (1994) A marker of oxyradical-mediated DNA damage (8-hydroxy-2'deoxyguanosine) is increased in nigro-striatum of Parkinson's disease brain. Neurodegeneration 3: 197–204

Schapira AHV, Cooper JM, Dexter DT, Clark JB, Jenner P, Marsden CD (1990a) Mitochondrial complex I deficiency in Parkinson's disease. J Neurochem 54: 823–827

Schapira AHV, Mann VM, Cooper JM, Dexter DT, Daniel SE, Jenner P, Clark JB, Marsden CD (1990b) Anatomic and disease specificity of NADH CoQ$_1$ reductase (complex I) deficiency in Parkinson's disease. J Neurochem 55: 2142–2145

Sian J, Dexter DT, Lees AJ, Daniel S, Agid Y, Javoy-Agid F, Jenner P, Marsden CD (1994) Alterations in glutathione levels in Parkinson's disease and other neurodegenerative disorders affecting the basal ganglia. Ann Neurol 36: 348–355

Sian J, Dexter DT, Lees AJ, Daniel S, Jenner P, Marsden CD (1994) Glutathione-related enzymes in brain in Parkinson's disease. Ann Neurol 36: 356–361

Sofic E, Reiderer P, Heinsen, Beckmann H, Reynolds GP, Hebenstreit G, Youdim MB (1988) Increased iron (III) and total iron content in post mortem substantia nigra of parkinsonian brain. J Neural Transm 74: 199–205

Authors' address: Dr. A. D. Owen, Pharmacology Group, Biomedical Sciences Division, King's College, Manresa Road, London SW3 6LX, United Kingdom

Treatment of behavioural disturbances in Parkinson's disease

F. Valldeoriola, F. A. Nobbe*, and E. Tolosa

Parkinson's Disease and Movement Disorders Unit, Neurology Service, Hospital Clínic i Provincial de Barcelona, Institut Pi Sunyer, University of Barcelona, Spain

Summary. Behavioural disorders in Parkinson's disease can grossly be subdivided in primary disturbances and those which are related to drug treatment.

Depression and anxiety are a common feature in parkinsonian patients. Both occur independently of drug treatment. In general, most current antidepressive and anxiolytic drugs could be administered in Parkinson's disease with the same precautions as in the normal population. However, in single case reports modern serotonin reuptake blockers in Parkinson's disease have been accused to worsen parkinsonian motor condition. Combinations of serotonin reuptake inhibitors with MAO-inhibitors like selegiline should be used with caution.

In the case of cognitive decline firstly an underlying depression should be disclosed or if existent be treated. Depression seems to be the single most important factor associated with the severity of dementia and early antidepressant treatment seems to decrease cognitive decline in depressed parkinsonian patients. Anticholinergic medications should be discontinued since they may cause mental side effects.

Sleep disorders in Parkinson's disease are mainly caused by nocturnal akinesia, which causes sleep fragmentation or altered dreaming and nightmares, which might be a side-effect of dopaminergic treatment. In the first case the administration of a controlled release preparation of levodopa at bedtime may be indicated. If the sleep disorder is considered to be due to dopaminergic medication, a reduction of long-term acting agents like modern dopamine agonists and controlled-release levodopa should be considered.

In severe psychotic states related to drug treatment antiparkinsonian therapy must be carefully analysed and, if possible, reduced. If motor condition worsens and/or psychiatric symptoms do not improve, initiation with "atypical" neuroleptics like clozapine is indicated. The pharmacological and clinical properties of new antipsychotic drugs that can be used in Parkinson's disease are revised.

*Supported by a Feodor-Lynen Fellowship from the Alexander von Humboldt-Stiftung, Bonn, Federal Republic of Germany

Table 1. Most common behavioural disturbances encountered in PD

— Depression
— Anxiety
— Cognitive impairment and dementia
— Sleep disorders
  • Sleep fragmentation
  • Daytime napping
  • Insomnia
  • Nightmares, altered dreaming
  • Drug-induced drowsiness
— Drug-induced psychiatric states
  • With clear sensorium (e.g. delusional state, hallucinations, paranoid ideations)
  • With clouded sensorium (e.g. delirium, confusion)

Introduction

Behavioural disturbances in Parkinson's disease (PD) are a common source of disability to both patients and their families but there is considerable controversy regarding their frequency and their neuropathological and neurochemical bases. Since they are so common, they must be clearly recognized and proper management instituted. The most frequent behavioural disturbances encountered in patients with PD are shown in Table 1. The main clinical characteristics, pathophysiology, and pharmacological treatment of these behaviours will be discussed in this chapter.

Depression in Parkinson's disease

General considerations

Depression is a frequent feature in PD, and has been reported with a widely variable prevalence ranging from 20 to 90% (Mindham, 1970; Celesia and Wanamaker, 1972; Mayeux et al., 1981; Gotham et al., 1986; Santamaria et al., 1986; Starkstein et al., 1990; Tison et al., 1995). The prevalence of depression in PD patients is higher than in age-matched control groups and may start years before the first motor symptoms (Santamaria et al., 1986). There is an insignificant or only weak association of depression with the severity of motor symptoms (Brown and Jahnshahi, 1995). Two studies demonstrated a non-linear relationship between depression and motor symptoms (Celesia and Wanamaker, 1972; Starkstein et al., 1990). Thus, depression in PD does not seem to be a consequence of progressive physical impairment since it even may antedate recognizable parkinsonian motor symptoms. The observation that depression occurs more frequently in PD patients with early onset of motor symptoms (Santamaria et al., 1986; Tison et al., 1995) could not be confirmed consistently (Hietanen and Teraevaeinen, 1988; Jankovic et al., 1990).

Clinically, depression in PD shares many features seen in primary affective depressive disorders. According to the DSM-IV criteria for the diagnosis of depression (American Psychiatric Association, 1994) one of the symptoms must be anhedonia or depressed mood for more than two weeks and, in addition, at least five of the following symptoms must be present: significant weight change; insomnia or hypersomnia; agitation or psychomotor retardation; fatigue or loss of energy; feelings of worthlessness or inappropriate guilt; decreased concentration and indecisiveness; and recurrent thoughts of death or suicidal ideation. Many PD patients without depression present one or more of the preceding symptoms. But these symptoms may be related to the motor disability produced by the disease rather than to an affective disorder. Rapid weight loss may be due to refractory rigidity. Also, insomnia is a common complaint in non-depressed PD patients. Hypomimia and bradykinesia may be misdiagnosed as fatigue or psychomotor retardation. Thus, weight change, sleep disorders, motor retardation and fatigue should be viewed with caution as diagnostic criteria for depression in PD or excluded altogether. For the assessment of current depression in PD various standardized depression scales are frequently used, i.e. Beck depression index (Beck, 1961) and Hamilton scale (Hamilton, 1960).

Selective hypometabolism in the bilateral caudate, orbital-inferior frontal and anterior temporal regions (Mayberg et al., 1990), and cingulate cortex and medial frontal cortex, respectively, (Ring et al., 1994) have been suggested as the pathological correlates for depression in PD (for review see Mayberg and Solomon, 1995). Similar alterations occur in patients with post-stroke depression (Robinson, 1988) and primary affective disorders (Baxter et al., 1989). Furthermore, dopaminergic cell loss in the ventral tegmental area is supposed to be positively correlated with depression in PD, as these neurons normally function as part of the brain's positive affective system (German et al., 1989).

Treatment and pharmacology

Once diagnosed, depression in PD should be treated promptly because of its clinical implications and the repercussions for a patient's quality of life. Furthermore, there is evidence for an association between depression and dementia in PD (Tröster et al., 1995; Starkstein et al., 1989) and early treatment of depressed PD patients may result in a less severe cognitive decline compared to those not treated (Starkstein et al., 1990). Early reports showed that levodopa not only improved motor symptoms but also exerted a positive influence on affective symptoms (Barbeau, 1969; Yahr et al., 1969; Celesia and Wanamaker, 1972). More recent studies have shown that levodopa treatment seems to have only a short beneficial effect on depression in PD presumably due to the amelioration of motor symptoms, since most patients with depression continue to be depressed one to six years after starting levodopa therapy (Shaw et al., 1980; see for review Santamaria and Tolosa, 1992). Consequently, most patients should be treated with antidepressants and the treating

Table 2. Common side-effects related to tricyclic antidepressants

Anticholinergic effects — Dry mouth — Constipation — Urinary retention — Blurred vision — Increased intraocular pressure	Anti H-1 effects — Excessive sedation
Anti-adrenergic effects — Postural hypotension — Tachycardia — Quinidine-like effect — Intraventricular conduction blockage	Miscellaneous — Ataxia — Nightmares — Disorientation — Insomnia — Parestesia — Peripheral neuropathy — Anxiety — Tinnitus
Related to abrupt withdrawal of the drug — Cephalalgia — Nausea — Dizziness — Fatigue — Sleep disturbances — Restlessness	Other effects — Allergic reactions — Diarrhoea, nausea — Galactorrhea — Libido disturbances — Ginaecomastia — Weight gain

Table 3. Side-effects related to serotonin reuptake blockers

Common side-effects
— Anxiety (occur in 10–15% of patients)
— Insomnia (occur in 10–30% of patients)
— Restlessness
— Weight loss, anorexia
— Somnolence or fatigue
— Postural tremor
— Light-headedness
— Diarrhoea and gastrointestinal disturbances
— Allergic reactions

Uncommon side-effects
— Worsening of motor symptoms in PD patients (?)
— Headache
— Akathisia
— Libido disturbances in men
— Dry mouth sensation (less common than in tricyclics)
— Increased sweating
— Hyponatremia (rare)

physician must be familiar with the clinical indications and side-effects of tricyclic antidepressants and serotonin reuptake inhibitors. The use of the classical non-selective monoamine oxidase inhibitors, and the combination of these drugs with other antidepressants should be avoided because of the high prevalence of side-effects (Feighner et al., 1990). Because the overall efficacy of most antidepressants is similar, the clinical usefulness of a given drug often depends on the presence or absence of particularly troublesome side-effects.

The serotonin reuptake inhibitors (i.e. fluoxetine, paroxetine) have advantages over the tricyclics regarding the side-effects spectrum. The anticholinergic features, typically associated with tricyclics, such as dry mouth and constipation, are less frequent with these drugs. Postural hypotension and sedation are infrequent but other serious adverse reactions have been reported with these drugs. The common side-effects of tricyclic antidepressants and serotonin reuptake inhibitors are shown in Tables 2 and 3.

For a severely apathetic PD patient, a stimulating antidepressant such as fluvoxamine, fluoxetine, sertraline or paroxetine should be prescribed. Early reports on the use of serotonin reuptake inhibitors in PD patients showed an improvement of mood and affective symptoms (McCance-Katz et al., 1992), but, to our knowledge, there are no controlled trials of any group of drugs in patients with PD. There have also been recent reports concerning the use of serotonin reuptake inhibitors in PD, that have pointed out the possibility of a worsening of motor symptoms during the treatment with fluoxetine (Bouchard et al., 1989; Jansen Steur, 1993) and paroxetine (Jiménez-Jiménez et al., 1994) which improved after withdrawal of the drug. Fluoxetine has also been associated with the induction of akathisia (Lipinski et al., 1989). However, reports of extrapyramidal side-effects related to serotonin reuptake inhibitors are anecdotal; a recent study showed no alteration of motor condition in 14 PD patients treated with a daily dose of 20mg of fluoxetine (Montastruc et al., 1994). Another study with 23 PD patients treated with fluoxetine and deprenyl, showed no adverse effects (Waters, 1994). The described negative effect of serotonin reuptake blockers on motor symptoms in PD was suggested to be due to a decrease in dopamine turnover or to an increased dopamine blockade related to the enhanced activity of the serotonergic neurons of the raphé nuclei projecting to the substantia nigra (Bouchard et al., 1989). Nevertheless, fluoxetine did not show significant changes on the striatal dopamine metabolism in rats (Baldessarini et al., 1992), and evidence for dopamine blockade seems to be low in experimental studies with paroxetine (Thomas et al., 1987). Theoretically, the increase of presynaptic serotonin levels might be helpful in parkinsonian depressed patients, since there are several biochemical and neuropathological evidences showing that serotonin neurons are involved in the disease process and the underlying source of the behaviour disturbances. Primary degeneration of the cortico-meso-limbic dopamine neurons in patients with PD, for example, may lead to dysfunction of the orbitofrontal cortex, which secondarily affects serotonergic neurons in the dorsal raphé resulting in impaired serotonin neurotransmission (Cowen, 1993; Mayberg and Solomon, 1995). Also, the brain content of serotonin in PD is globally reduced and the levels of the main metabolite of serotonin, 5-hydroxyindolacetic acid, are diminished in a subgroup of PD patients prone to depression (Mayeux et al., 1984, 1986). An average loss of more than 50% of Nissl-stained large neurons in the dorsal raphe nucleus and a similar loss of serotonin neurons in the median raphe nucleus of PD patients have been reported (Halliday et al., 1990). It has also been shown that serotonergic terminals projecting into the basal ganglia are affected in PD (Chinaglia et al., 1993). Furthermore, autoradiographic studies have shown that serotonergic reuptake sites are decreased by 40–50% in PD (D'Amato et al., 1987).

In conclusion, very few single case reports have suggested that serotonin reuptake blockers worsen the motor condition in PD. Moreover, these drugs produce milder side-effects than tricyclic antidepressants, especially in the elderly population, and they should be regarded as the first choice drugs in the treatment of depression in PD. They should be used cautiously until controlled studies are carried out. The development of new serotonin reuptake blockers which also are 5-HT2 receptor antagonists, such as nefazodone (Eison et al., 1990; Fontaine, 1993), with combined antidepressive and anxiolytic properties (Feighner et al., 1989) and a reduced side-effects spectrum, will probably be of interest in the treatment of depressed PD patients in the future.

Anxiety

General considerations

Anxiety disorders like depression are common in PD patients. Stein and colleagues (1990) found anxiety disorders in 38% of patients with PD, not related to the motor impairment or treatment with dopamine agonists. Many parkinsonian patients feel excessive fear, motor tension, and autonomic hyperactivity. Social phobias, fear of walking or falling and avoidance of crowds, are also common features. Motor tension in combination with anxiety could be confused with akathisia, a frequent feature of PD (Lang and Johnson, 1987) or could be related to the use of neuroleptic drugs; in such cases careful clinical history-taking and neurological examination are paramount in order to establish the correct diagnosis. The necessity to be in constant motion, the incapacity to sit or stand still and a drive to pace up and down, are typical features of akathisia.

Treatment

There is a long list of drugs that can be used in the management of anxiety symptoms in PD. Benzodiazepines have a widespread use in the general population as well as in PD. Lorazepam is an useful agent since it is relatively short acting (half-life 10 to 15 hours) and well tolerated in older patients. It must be taken twice or three times daily, but dosage should be increased slowly in order to avoid excessive sedation or other side-effects. Other benzodiazepines such as diazepam or chlordiazepoxide, with a longer half-life, can also be used, but they are poorly tolerated by older patients and patients with impaired renal function. When discontinuing their use, these drugs must be tapered over a two-week period. The long-acting benzodiazepines (i.e. diazepam and chlordiazepoxide) have the advantage that they can be prescribed once daily, usually at bedtime, because of their accumulation in the central nervous system and less risk of withdrawal symptoms.

Barbiturates are not commonly used in PD patients because of sedative side-effects.

Drugs which have effects on the autonomic nervous system, such as diphenhydramine or hydroxyzine, may also be used in patients with symptoms of autonomic dysfunction such as wet and cold hands, tachycardia, tachypnea, dry mouth, diarrhoea and parestesia in hands or feet. Beta blockers, which can be used for the treatment of postural tremor in PD, have a beneficial effect on anxiety-related palpitations. However, they should not be prescribed in patients with a heart block or impaired respiratory function.

Another drug with anxiolytic properties is buspirone chlorhydrate. It is not chemically related to barbiturates or benzodiazepines and has no anticonvulsive or myorelaxant properties. Although its mechanism of action is not completely understood, a partial agonism at serotonin type 1A receptors has been demonstrated (Andrade and Nicoll, 1987). It has weak dopaminergic agonistic and antagonistic activity, as well (Riblet et al., 1982; Algeri et al., 1988). The antagonistic dopaminergic activity could theoretically produce parkinsonism, akathisia or other neuroleptic-related side-effects. Nevertheless, in controlled studies with normal subjects these effects were not observed, except for motor restlessness that was noted in a few patients. Once administered, the drug is rapidly distributed. Its metabolite, 1-primidinil-piperazine, has about four-fold less activity. Buspirone does not have the sedative side-effects of benzodiazepines and seems to be effective in reducing aggressivity (Gedye et al., 1991). The recommended dose is 5 mg three times daily or 10 mg twice daily and the maximal dose should not exceed 60 mg daily. Clinical studies showed that buspirone is not as effective as benzodiazepines for anxiety symptoms but the clinical characteristics of likely responders and non-responders are still to be delineated (Gedye, 1991).

Dementia and cognitive deficits

General considerations

Cognitive decline in Parkinson's disease is now recognized as a frequent and important feature of the illness. However, in the first description of the disease James Parkinson reported that "the senses and the intellect remained uninjured" (Parkinson, 1817). This view of PD as a neurodegenerative process sparing the mental functions was dominant until this century. In the last decade a large number of studies were carried out to detect the frequency and the nature of the cognitive changes in Parkinson's disease. It is important for the patient, his family and the attending physician to know about the features and the time-course of a possible mental deterioration and its relationship to medication and concomitant depression.

Two different recent studies revealed an overall prevalence of dementia in PD patients of 41.3% and 17.6% (Mayeux et al., 1992; Tison et al., 1995). In a cohort of 82 PD patients the incidence of dementia was estimated as 47.6/1,000 person years of observation (Biggins et al., 1992); this means that in a 5-year period almost every fourth PD-patient becomes clinically demented. Most studies agree that age and/or late onset of parkinsonian symptoms are the main predictive factors for mental deterioration in PD (Hietanen and Terävainen, 1988; Jankovic et al., 1990; Mayeux et al., 1990, 1992; Biggins et al., 1992; Caparros-Lefebvre et al., 1995; Palazzini et al., 1995; Tison et al., 1995). Other studies have shown a relationship between dementia with age and institution-alization (Tison et al., 1995), with the underlying pathology (Rajput et al., 1992) and with the clinical presentation, the cognitive decline being more common in patients with postural instability and gait disorders than in those with tremor as the main symptom (McDermott et al., 1992). Also, isolated dystonic dyskinesia and worse motor response on levodopa have a negative predictive value on cognition (Caparros-Lefebvre et al., 1995). Several studies have shown that depression affects cognitive performance in PD (Starkstein et al., 1989; Stern et al., 1993; Tröster et al., 1995). Further, while on imipramine, depressed PD patients showed a less severe cognitive decline compared with those who were not treated (Starkstein et al., 1990). Depression seems to be the single most important factor associated with the severity of dementia in PD (Starkstein et al., 1989). Non-demented or depressed PD patients on medica-tion were found to be able to carry out visuospatial tests as rapidly as normal healthy controls (Duncombe et al., 1994).

The cognitive disorder which is associated with PD is commonly regarded as a "subcortical dementia" (Pirozzolo et al., 1993). The term was introduced by Albert et al. (1974) to characterize the mental decline pattern in progres-sive supranuclear palsy. It involves the following characteristic features: slow-ness of information processing, altered personality, forgetfulness, predominantly short-term memory, and difficulty in manipulating acquired knowledge. The absence of "cortical" deficits such as apraxia, agnosia and aphasia in PD-patients is the chief feature to distinguish PD-dementia from "cortical" dementias like in Alzheimer's disease.

For the clinical diagnosis of dementia related to PD the DSM criteria are widely used. According to the DSM-IV criteria there must be an impairment of short and long-term memory (criterion A.1.) and an impairment of at least one other "higher cortical function" (A.2.); i.e. aphasia, apraxia, agnosia, impaired abstract memory, impaired judgement, constructional difficulty or personality change. Criterion B requires interference with occupational or social functioning, a non-organic mental illness as the cause of dementia must be excluded (criterion C) and the demented state must be independent on delirium or an alteration in consciousness. (American Psychiatric Association, 1994.) The Mini-Mental State Examination (Folstein et al., 1975) is a wide-spread test of cognitive functions which may reveal possible concomitant mental deficits in PD and helps the physician to diagnose a suggested dementia and to assess the course of cognitive decline. However, the test is criticized because of its shortage of executive type tasks; i.e. conceptual-ization, initiation and perseveration (Mohr et al., 1995). Thus, a dementia scale, like the Mattis Dementia Rating scale (Mattis, 1976), which includes executive tasks seems to be more appropriate for the assessment of intellec-tual changes in PD. For the detection and the assessment of subtle cognitive

Table 4. Proposed neuropathological substrate of PD dementia

— Alzheimer-type pathology
— Cortical Lewy bodies and coexisting PD changes
— Nigrocortical degeneration
— Neuronal loss in the nucleus basalis of Meynert
— Lewy bodies in the raphe nuclei: impaired serotonergic function
— Lewy bodies in the locus coeruleus: impaired noradrenergic function

changes, more complex and specific neuropsychological tests need to be used. In the early stage of the disease subclinical cognitive deficits can be found. Several neuropsychological studies showed deficits in the following cognitive spheres. Planning and conceptualization, ability to profit from feedback and flexibility of thought in the solution of novel problems are subsumed as executive capacities, which are supposed to be a frontal lobe function. A deficit in execution and conceptualization was found to be a major feature of cognitive deficits in PD patients (Litvan et al., 1991; Martí and Tolosa, 1991; Owen et al., 1992; Palazzini et al., 1995; Tröster et al., 1995). Also, in the early stage of the disease subclinical frontal dysfunction was found and an aggravation of these symptoms by anticholinergic drugs has been suggested (Farina et al., 1994). Especially, an impairment of item recognition when a secondary task is required revealed a deficit in attentional resources (Cooper and Hagar, 1993). Finally, deficits in the visuospatial function are a well described finding accompanying PD (Brown and Marsden, 1986). Anticholinergic medication is reported to impair memory function in PD patients (Dubois et al., 1987; Sadeh et al., 1982; Cooper et al., 1992). The observation that short-term memory recovers at least partly after withdrawal of long-term anticholinergic therapy (van Herwaarden et al., 1993) could explain different results obtained in a retrospective analysis of the effects of anticholinergic medication on memory performance, which showed no evidence of memory deterioration due to anticholinergic drugs (Levin et al., 1992).

In a retrospective clinico-pathological study of 18 consecutive and unselected Parkinson's disease patients, Alzheimer-type pathology was supposed to be the major determinant of the cognitive decline in demented PD patients (Table 4). It was not possible to distinguish a specific pattern of cortical Lewy bodies in PD patients with and without dementia (De Vos et al., 1995). Cortical Lewy bodies are an observation in most PD cases and do not preclude a diagnosis of idiopathic PD (Hughes et al., 1992). On the other hand in a study on cortical Lewy bodies in Alzheimer's disease an association was found between the presence of cortical Lewy bodies and extrapyramidal symptoms (Kazee and Han, 1995). Other pathological substrates of dementia in PD have been proposed: somatostatin concentrations in the CSF were found to be significantly lower in demented versus non-demented PD patients (see Table 4, Martí and Tolosa, 1991).

Treatment

In the absence of definitive therapies and controlled clinical trials the management of cognitive impairment in PD involves non-specific symptomatic treatment and counselling of the patient and the family. If patients become severely demented or helpless, education and training of the family as well as nursing home care and outside support are fundamental.

Modification of the outcome of dementia and of the quality of life for the patient by anticholinesterase inhibitors, cholinergic agonists or neurotransmitter replacements is doubtful. The use of selective monoamino oxidase B inhibitors like deprenyl (selegiline) might have some application as putative neuroprotective properties (Jankovic et al., 1990; Parkinson Study Group, 1993). However, it has been shown that neither deprenyl nor tocopherol were effective on cognitive test performance in early untreated PD patients (Kieburtz et al., 1994). Piracetam was tested by Sano and co-workers (1990) but no improvement in cognitive symptoms was found. In another study, Fungfeld and colleagues (1989) demonstrated that phosphatidalserine increased the speed of background EEG activity in PD patients but no effect on cognitive function was noted. Promising results have been reported recently from a pilot study with famotidine suggesting a possible role of H2 antagonism in therapy of PD-related executive deficits (Molinari et al., 1995). The authors described statistically significant improvement in the performance of executive tasks after six weeks' treatment with famotidine, 40 mg/12 h. However, the selection criteria of PD patients in this study are unclear. Further studies will be needed to confirm the hypothetical benefit of H2 antagonists on cognitive decline in PD. Prevention of the development of psychiatric disturbances, such as psychosis or hallucinations, is a basic question since demented parkinsonian patients are more prone to develop these complications as a consequence of the drug therapy (Mohr et al., 1995). This might be due to changes in the cortical serotonin 5-HT2 receptors as demonstrated in receptor binding studies (Cheng et al., 1991). Other possible causes of dementia should be ruled out, such as metabolic or endocrine alterations capable of causing confusion or cognitive decline. Anticholinergic medications are likely to produce confusion and disorientation, and aggravate memory performance, so they should be tapered and discontinued in demented patients (Sadeh et al., 1982; Koller, 1984). These drugs, if needed, should be reserved for the earlier phases of the disease or for younger patients because mental side-effects are more likely to occur in the elderly. It is essential to avoid dangerous combinations of drugs such as monoamine oxidase inhibitors and tricyclic antidepressants or serotonin reuptake blockers because of their ability to cause hallucinations or delusions, especially in older patients.

In the early days of dopaminergic therapy, it was believed that levodopa might improve intellectual function in PD (Meier and Martin, 1970; Beardsley and Puletti, 1971; Donelly and Chase, 1972). Then, levodopa was shown to improve memory (Donelly and Chase, 1972; Halgin et al., 1977), concept formation (Bowen et al., 1975) and perceptual skills (Harnel and Riklan, 1975). These effects, however, might simply reflect a secondary effect of improved attention or vigilance. Subsequently, it was found that the relation-ship between intellectual function and the levodopa use was influenced by the duration of treatment (Rinne, 1983). In this study, improvement of intellec-tual performance was seen in the first year after levodopa therapy initiation, but after this initial period, a progressive return to baseline cognitive function was observed despite the continuation of levodopa therapy. Memory impair-ment was demonstrated in patients with PD in which plasma levels of dopam-ine were associated with the wearing-off phenomenon. This means that the ability to recall was impaired in these patients when dopamine plasma levels varied up or down compared with situations in which dopamine levels were stable (Huber et al., 1987, 1989). Thus, at the present moment, there are no reliable studies showing that levodopa has a positive impact on cognition.

The extent to which surgical procedures such as thalamotomy, pallidotomy and fetal tissue transplantation, alter the progression and symptomotology of parkinsonian symptoms, including cognitive symptoms, remains unknown (Spencer et al., 1992; Narabayashi, 1996; Dogali et al., 1996) except in patients with juvenile parkinsonism (Tasker et al., 1996) in which dementia is extremely uncommon. These surgical procedures do not seem to have a negative effect on cognition in non-demented parkinsonian patients, but long-term follow-up studies are still needed (Dogali et al., 1995; Lozano et al., 1995).

Sleep disturbances

General considerations

In addition to the loss of dopamine, there are other neurotransmitters involved by the disease process in PD. Serotonin levels are reduced in the brain and the metabolite of serotonin, 5-hydroxyindolacetic acid concentra tion is low in the cerebrospinal fluid of PD patients (Mayeux et al., 1984, 1986). Norepinephrine levels and tyrosine hydroxylase and glutamic acid decarboxylase activity are also decreased (Fahn et al., 1971; Javoy-Agid et al., 1981). A fully developed neurochemical model of sleep has yet to be discov-ered. However it is known that the serotonergic system has a regulatory function in slow-wave sleep (Mouret et al., 1968; Jouvet, 1969) and the timing of the initial rapid eye movement (REM) cycle appears to be dependent on cholinergic activity. Taking into account that there is a reduction of serotonin and the cholinergic activity is higher than in normal subjects, sleep distur-bances in PD could be explained on a specific neurochemical basis. The effect of dopamine depletion on sleep is more difficult to explain since dopamine receptor blockers or dopamine depleting agents alter multiple biochemical systems and do not seem to have a specific effect on sleep. Nevertheless, direct dopamine agonists, such as apomorphine, have a suppression effect on REM sleep and slow-wave sleep; apomorphine induces sleep at low doses and reduces total sleep time at higher doses (Kafi et al., 1976; Cianchetti et al., 1985).

Treatment

In untreated parkinsonian patients a common complaint is sleep fragmentation. Sleep disturbances seem to be related more to the underlying process
than to the medication and insomnia in PD is often associated with depression
or anxiety. Insomnia can produce excessive daytime napping. In such patients,
initiation of levodopa therapy seems to improve nocturnal sleep, suggesting a
direct relationship between sleep disturbances and motor symptoms
(Nausieda, 1992). In the dopamine-treated PD population, sleep disturbances
are more common but this might be a bias produced by an older population
and a longer duration of the disease rather than an effect of dopaminergic
medication. Altered dreaming, nightmares and vocalizations related to
dopaminergic medication, generally precede hallucinations or psychosis.
Some patients, in the initial phase of treatment with dopaminergic drugs,
complain about insomnia, a side-effect also observed in normal subjects
(Fram et al., 1970; Nausieda et al., 1982). This might be related to the amphet-
amine-like effects of dopaminergic drugs. This effect is mostly transient, thus,
if patients report insomnia in the first days after initiation of levodopa or
levodopa agonists therapy, there is no reason to start sedative medication
except if the symptoms are a major concern. In such cases a short-acting
benzodiazepine, such as alprazolam, should be used and tapered when the
complaint disappears. Some PD patients become drowsy after each dose of
levodopa; this can be resolved by adjusting the amount and the timing of each
individual dose of levodopa. Stimulating drugs during the day should not be
used except for the exceptional non-responsive cases for whom daytime
drowsiness is a major complaint; caffeine and, in rare cases, amphetamines,
are recommended.

Improved motor function in the morning is suggested to be the result of
storage of dopamine in the nigral neuronal terminals during sleep in PD
treated patients (Marsden et al., 1982). This could be one factor for the
improved motor function in the morning compared to the rest of the day
observed in many PD patients. Factor and colleagues (1990) believe that sleep
fragmentation is an effect of motor fluctuations in treated parkinsonian patients and did not find any relationship between sleep itself and morning
motor function, but total doses and timing of antiparkinsonian medication
were not studied.

A polysomnograph and video study of 12 PD patients with sleep disturbance showed that sustained-release levodopa was more effective than standard levodopa in improving sleep efficiency, consolidation and subjective
evaluation. This improvement in sleep was not associated with changes of
motor function (Nausieda et al., 1994).

The treatment of sleep fragmentation in PD is difficult. The short-acting
benzodiazepines may help initially but tolerance develops. A carefully noted
clinical history should be directed to reveal the eventual co-existence of other
disturbances such as depression, anxiety or dopaminergic-related side-effects
(nightmares, hallucinations, etc.). It is important to detect the existence of
undesirable awakenings produced by nocturnal motor difficulties or nocturia.
There is still some controversy whether awakenings are produced by the

parkinsonian symptoms alone or reflect the sleep fragmentation observed both treated and untreated PD patients. Consequently, dopaminergic medication should be increased at bedtime, but other approaches aimed to consolidate nocturnal sleep may also improve symptoms. In cases in which benzodiazepines are not effective, tricyclic antidepressants might be used but there is a risk of precipitating hallucinations or confusion in patients with associated parasomnias (Nausieda et al., 1983) a phenomenon relatively frequent in the early stages of levodopa therapy (Scharf et al., 1978; Moskovitz et al., 1978). Because of its sedative properties, clozapine, an atypical neuroleptic, may be of interest in the treatment of sleep disturbances in PD (Friedman and Lannon, 1989).

Drug induced psychotic states

General considerations

Psychosis in untreated Parkinson's disease patients is rare (Factor et al., 1995) but in patients treated with dopaminergic drugs, psychiatric symptoms are much more common and may affect 10–50% of PD patients (Goodwin, 1971; Moskovitz, 1978; Rinne, 1983; Wilson and Smith, 1989). Although psychiatric states can be induced or aggravated by other drugs, such as amantadine, selegiline, anticholinergics and even propranolol, the most common drugs associated with these side-effects are dopamine agonists. It is well known that the substances which are structurally similar to dopamine, such as lysergic acid diethylamide, mescaline and amphetamines are prone to cause hallucinations and other psychiatric symptoms (Yaryura-Tobias et al., 1970). Administration of levodopa causes an increased turn-over of serotonin attributed to the overstimulation of central serotonergic 5HT-3 receptors, and this has been proposed to be the mechanism responsible for levodopa induced hallucinations (Melamed et al., 1993).

Drug-induced psychoses may occur in isolation or in association with a toxic confusional state. Benign organic hallucinosis with a normal state of consciousness is the most common mental side-effect associated with dopaminergic therapy (Factor et al., 1995). Drug-induced psychosis with affected sensorium is considered to be the most troublesome psychiatric symptom and it is a frequent reason for nursing home placement in PD patients (Goetz, 1993). The incidence of mental side-effects is higher in demented patients, in patients exposed to higher doses of dopamine agonists, in patients with premorbid psychiatric illness, and in the elderly. The combination of levodopa with other PD medications is much more likely to cause confusional states than when levodopa is used alone (for review see Factor et al., 1995).

Treatment

Once psychosis is diagnosed, a complete general clinical and laboratory examination should be carried out in order to eliminate meta-

Table 5. Clinical management of psychosis in PD

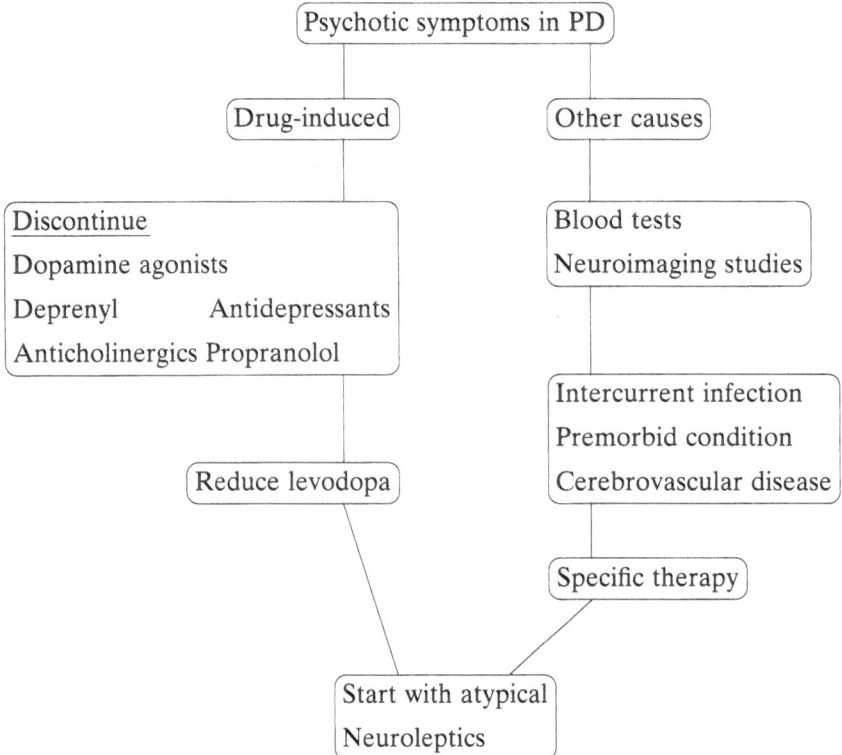

bolic, endocrine or other general or neurologic diseases causing mental disturbances.

Once other intercurrent underlying conditions have been ruled out, psychiatric illness should be considered to be related to dopaminergic medication. Dopamine agonists such as bromocriptine, pergolide, cabergoline, ropinirole, lisuride, etc., are more likely to produce psychiatric symptoms than levodopa. In the case of combination therapy with dopamine agonists, treatment should be reduced to levodopa monotherapy. Antiparkinsonian drugs like deprenyl or anticholinergics should be discontinued. If the condition of the patient is not improved or the psychiatric symptoms produce significant morbidity, the dose of levodopa should be reduced. It may be necessary to admit the patient to hospital in order to control and prevent complications from lowering dopaminergic therapy. Psychiatric symptoms may be persistent for many days despite drug reduction, meanwhile the motor condition of the patient gradually worsens. Therefore, it is reasonable to maintain levodopa therapy or reduce the dosage very slightly. If there is no improvement, the initiation of antipsychotic drugs is paramount. An algorithm for clinical management of psychosis in PD is summarized in Table 5. Different drugs were studied for the treatment of psychiatric symptoms in PD. The serotonergic precursor tryptophan (Beasley et al., 1980), methylsergide, and physostigmine

(Friedman, 1991) are ineffective; electroconvulsive therapy has also been used, but the practicality of this method has been questioned (Hurwitz et al., 1988). In the past, standard or "classical" neuroleptics were used, but the marked exacerbation of parkinsonian symptoms is a major drawback (Lipper, 1976) and the greater risk for tardive dyskinesia in the elderly population severely restrict the use of these drugs. The so-called "atypical antipsychotics" and some compounds that interact with the serotonergic system have emerged recently as therapeutic alternative for PD patients with thought disorders.

Atypical antipsychotic drugs

Clozapine

Clozapine (N-methyl-pyperazinyl-dibenzodiazepine) is an atypical neuroleptic and emerged as a good alternative in the treatment of psychiatric disorders in PD patients (Tolosa et al., 1994). This drug was developed over 30 years ago and used in psychotic patients refractory to standard neuroleptics. However, it was abandoned because of its haematological side-effects. This dibenzodiazepine derivative has strong antipsychotic and sedative properties and lesser extrapyramidal side-effects than standard neuroleptics like butyrophenones and phenothiazines. It does not increase prolactin secretion (Meltzer et al., 1989a). In the parkinsonian rat model, clozapine does not block behaviour arousal induced by dopamine agonists such as apomorphine and amphetamine, and unlike other neuroleptics, does not increase the sensitivity or the number of dopamine receptors in the basal ganglia (Bürki et al., 1975; Rupniak et al., 1985; Moore and Kenyon, 1994).

The pharmacokinetic properties of clozapine include a marked interpersonal variability in the plasma levels of the drug depending on age, gender, and weight. The correlation between clinical efficacy and the plasma levels is weak. The metabolism is mainly hepatic and protein binding is very high. The major concentrations of the drug are obtained in the brain, kidneys, lungs and liver.

The pharmacodynamic characteristics of clozapine include selective antidopaminergic, anticholinergic (muscarinic), antiadrenergic (1 and 2), antiserotonergic (5-HT1, 5-HT2, 5-HT3, 5-HT6 and 5-HT7) and antihistaminergic (H1) properties (Cohen et al., 1979; Richelson et al., 1984; Meltzer et al., 1989b). Several pharmacological properties differentiate this drug from the classic neuroleptics which could explain the lesser extrapyramidal side-effects of clozapine. Positron emission tomography studies have demonstrated that clozapine has a relatively high occupancy of D1 receptors in the striatum and relatively low D2 affinity when compared to standard neuroleptics (Farde et al., 1989, 1992; Pilowski et al., 1992). Blockade of D2 receptors have been proposed as the basic mechanism producing tardive dyskinesia and parkinsonism associated with traditional neuroleptic (Walker et al., 1988). Other explanations include its lack of binding to the haloperidol-sensitive receptors (Walker et al., 1988) and the suggested preferential action of clozapine on dopamine σ receptor blockade in the limbic system rather than in the basal ganglia (Borison et al., 1981; Anderson et al., 1988). Recently, it has been shown that clozapine and its congener clorotepine have high affinity for the D3 and D4 dopamine receptor (Van Tol et al., 1991; Seeman, 1992), mainly located in the limbic system, suggesting that clozapine could produce its antipsychotic activity primarily by blocking these receptors in the limbic areas (Sibley, 1992). The polymorphism of D4 receptor does not affect clozapine affinity (Shaikh et al., 1993). The pathophysiologic bases of schizophrenia are supposed to be related to the D4 receptor (Iversen, 1993; Seeman et al., 1993). The clozapine antagonism to 5-HT2 receptors was suggested to be the basis for its efficacy in the treatment of levodopa-induced psychosis (Meltzer et al., 1995). A summary of the differential properties of clozapine compared to standard neuroleptics is shown in Table 6.

There are several studies using clozapine to treat psychosis in parkinsonian patients. Friedman and Lannon (1989) reported a non-blinded study on six patients in whom clozapine improved mental disorders without producing worsening of parkinsonian symptoms. Three of these patients presented improvement in motor condition. In a quantitative study, Factor and Brown (1992) found statistically significant improvement of psychosis in seven out of eight patients after four weeks' treatment with clozapine. In an open pilot trial using this drug and fluperlapine, improvement of psychosis was found in 11 out of 13 patients, with no increase in parkinsonian disability

Table 6. Main differential characteristics of the atypical neuroleptic clozapine in relation to classic neuroleptics

Pharmacodynamic characteristics
— No blockade of stereotyped locomotion induced by dopamine agonists in rats
— No increase of prolactin secretion
— No upregulation of dopamine receptors in the basal ganglia
— High affinity to D1 receptors
— Low affinity to D2 receptors
— High affinity for the D3 and D4 dopamine receptors
— Preferential blockade of dopamine receptors in the limbic system
— Lack of binding to the haloperidol-sensitive receptors
— Antagonism to 5-HT1, 5-HT2, 5-HT3, 5-HT6 and 5-HT7 receptors

Clinical characteristics
— Serious idiosyncratic haematological side-effects in 1–3% of patients
— Lack of extrapyramidal side-effects at low doses
— Higher tendency to produce seizures
— Improvement of motor condition in inter-dose "off" periods in PD
— Applicability in the treatment of akathisia, tremor and dyskinesia

(Scholz and Dichgans, 1985). Pfeiffer et al. (1990) obtained good results in 14 out of 16 patients with low doses of clozapine and Ostergaard and Dupont (1988) and Lew et al. (1993) also reported good results. In contrast, in the unique double-blind study on clozapine in PD psychosis, Wolters et al. (1990) found an improvement in the psychiatric disorder but a deterioration of the motor condition in three out of six patients. The authors used doses of clozapine, 75–250 mg daily, while others used doses between 6.5 and 100 mg daily. Although far lower than the doses used in schizophrenia (Meltzer et al., 1995), the cause of the worsening of motor symptoms in these parkinsonian patients is probably due to the higher doses used. The lack of other double-blind studies with lower doses might be caused by the difficulty of maintaining the test conditions when clozapine produces strong sedation.

Extrapyramidal side-effects were seen only in about 4% of schizophrenic patients with high dose levels (Gerlach and Peackock, 1994). Sedation is one of the major side-effects of clozapine, reported in 20–30% of patients (Juul-Povlsen et al., 1985; Pfeiffer et al., 1990). Other adverse effects include dry mouth, orthostatic hypotension and hypersalivation (Jansen, 1994). EEG alterations and seizures (Simpson and Cooper, 1978) seem to be related to plasma levels and have a little higher incidence than with standard neuroleptics (Haring et al., 1994). Seizures have been described even at low doses (Thomas and Goudeman, 1992). Weight gain (Cohen et al., 1990), constipation, diarrhoea and other gastrointestinal disturbances are relatively frequent (Launer, 1992). Uncommon side-effects include prolonged post-ictal encephalopathy, described in two patients (Karper et al., 1992); neuroleptic malignant syndrome (Pope et al., 1986; Müller et al., 1988); dizziness and sleep disturbances (Baldessarini and Frankenburg, 1991); and increase of liver enzymes (Naber et al., 1992).

The most serious adverse effect of clozapine are haematological complications. Mild to moderate granulocytopenia may be expected in 1.5–3% of patients during treatment (Krupp and Barnes, 1992; Gerson et al., 1993); a severe agranulocytosis is expected to occur in 0.8% of patients treated for one year. The risk of serious complications can be predicted and minimized by regular white blood cell monitoring. This drug is contraindicated in patients with a history of blood dyscrasia and should be never be combined with other drugs with a potential to depress the bone marrow, like ticlopidine, carbamazepine and captopril. The bone marrow toxicity is an idiosyncratic response to clozapine and it is not dose-related. The risk is higher in the elderly, in men, and in the first three or four months after the onset of the therapy. Toxic and immunologic mechanisms have been proposed and a possible ethnic predisposition has been shown in Jewish and Finnish patients (Idänpään-Heikkilä et al., 1975; Krupp and Barnes, 1992).

Clozapine should be started at very low doses of 6.5–12.5 mg daily and increased progressively but keeping the dose to a maximum of 75 mg daily under strict haematological control. Periodic white-cell count must be done during therapy and the drug must be immediately discontinued when granulocytopenia is detected. The sedative effect of clozapine has been clinically used in the treatment of sleep disturbances in PD patients (Friedman and Lannon, 1989). Clozapine was also reported to improve motor function in PD patients. This was confirmed in a trial performed by Gómez Arévalo and Gershanik (1993) in six PD patients. Parkinsonian disability scores were reduced after treatment with clozapine, especially in the interdose "off" periods. These patients presented the well-known phenomenon seen in PD of greater motor disability during the interdose "off" periods when compared to that observed during the early morning akinesia period. This phenomenon has been attributed to a possible inhibitory action of levodopa and D2 agonists when plasma concentrations are at sub-threshold levels (Nutt et al., 1988). Different affinities to the ambivalent pre- and postsynaptic D2 receptor subtypes were proposed for its stimulatory and inhibitory effects on motor behaviour. Clozapine may have a greater affinity for these inhibitory D2 receptors and, therefore, reduces the inhibitory effect of low levels of levodopa. As pointed out by Gómez Arévalo and Gershanik, this fact correlates well with the experimental observation in mice that low doses of D2 antagonists are necessary to block the inhibitory action of low-level D2 stimulation (Rubinstein et al., 1988).

The efficacy of clozapine in the treatment of drug-induced psychosis in PD has been demonstrated. It has also been proposed as a possible preventive therapy to avoid mental symptoms during the increase of dopaminergic medication to improve parkinsonian disability (Factor and Brown, 1992). The activity of the drug, however, seems to decline in the second year of treatment, perhaps due to a progression of dementia or to the increase of levodopa dose (Factor et al., 1994).

Risperidone

Risperidone is a benzisoxazole derivative with a potent antagonism to 5-HT2 and to a lesser degree to dopamine D2 receptors (Livingston et al., 1994). In fact it was demonstrated that risperidone is at least as potent as the specific 5-HT2 receptor antagonist ritanserin. The affinity of risperidone to D4 and D3 receptors is five and nine times lower, respectively, than to D2 receptors. It interacts with 1 adrenergic and H1 receptors but is devoid of significant anticholinergic activity (Megens et al., 1994) and, unlike clozapine, it has little affinity to D1 receptors (Meltzer, 1993).

Risperidone is rapidly absorbed in the gastrointestinal tract with plasma peak levels about two hours after administration and 24 hours of action. The main metabolite is 9-hydroxyrisperidone. Half-life of the drug is longer in older patients and in patients with renal insufficiency. Risperidone has 88% plasma binding and is rapidly distributed. Brain concentrations of risperidone and 9-hydroxyrisperidone are 10–18 times higher in the frontal cortex and the striatum with high 5-HT2 and D2 receptors density, than in the cerebellum (Van Beijsterveldt et al., 1994).

This compound was tested in several trials in schizophrenic patients showing an important improvement of positive symptoms and a superior activity than haloperidol on negative symptoms (Borison et al., 1995). Rarely extrapyramidal side-effects were observed (Borison et al., 1992; Claus et al., 1992;

Chouinard et al., 1993) suggesting that the mixed antiserotonergic and antidopaminergic effects in the brain are efficient in the reduction of psychotic symptoms without producing parkinsonism (Jansen et al., 1987; Meltzer et al., 1989). This drug, administered 4 mg/day seems to have similar antipsychotic efficacy than clozapine administered 400 mg/day, with similar low extrapyramidal side-effects but better toleration (Heinrich et al., 1991). In a open-label study on six parkinsonian patients with levodopa-induced hallucinations, Meco et al. (1994) reported disappearance of psychotic symptoms in three out of six patients treated with risperidone and significant improvement in the others. The motor condition did not worsen and the daily levodopa dose remained unchanged as well as the total "on" period time. On the other hand, Ford et al. treated six patients with this compound at slightly higher doses and found a clear worsening of motor symptoms and decline of daily living capacities (1994).

The most common, adverse experiences with this drug are slight postural hypotension, related to its α-adrenergic activity, hypersalivation, and weight-gain, a common complication of antipsychotic agents. Prolactin levels rise in a dose-related manner producing decreased libido, galactorrhea, gynecomastia and menstrual upset. Extrapyramidal side-effects and sedation are infrequent. There are no haematological complications described with risperidone.

Risperidone should be regarded as a promising drug to treat psychotic symptoms in PD due to its efficacy, demonstrated in schizophrenic patients. In the two studies performed in only few parkinsonian patients, the results regarding extrapyramidal side-effects were contradictory. Further studies are needed to test the clinical utility of this new drug in parkinsonian patients.

Olanzapine

Olanzapine is a thienobenzodiazepine derivative with a high affinity to 5-HT2, and dopamine D1 receptors, and also to D2, D4, muscarinic, α-1 adrenergic and H1 receptors with a similar activity profile to clozapine (Moore et al., 1992). In animal models olanzapine has greater activity at serotonergic than dopaminergic sites. It was demonstrated that this compound produces antagonism of apomorphine-induced climbing behaviour in rats, a test widely used to predict antipsychotic activity (Moore and Axton, 1988). Nevertheless, olanzapine produced cataleptic activity in rats only at high doses, an observation indicating a low tendency to produce extrapyramidal side-effects (Moore et al., 1992). In common with clozapine, electrophysiological studies demonstrated that chronic administration of olanzapine decreases firing activity of mesolimbic but not striatal dopaminergic cells. Prolactin elevation is significantly less frequent with olanzapine than with haloperidol.

A recent study on olanzapine (12.5–17.5 mg/day) with 335 schizophrenic patients, showed its superior efficacy over haloperidol on negative symptoms with fewer extrapyramidal side-effects (Borison et al., 1995). Haematological complications have not been seen in the 2,500 psychiatric patients studied so far. The most common adverse side-effects associated with olanzapine are dose-related drowsiness or sedation, liver enzyme elevation without clinical relevance, dry mouth, dizziness and weight-gain. There are no therapeutic controlled trials with olanzapine in dopaminergic psychosis reported to date. Such studies have been initiated and will hopefully reveal its clinical utility in the treatment of drug-induced psychosis in PD.

Other atypical antipsychotic drugs

Several newly-developed compounds that can be considered as atypical neuroleptics are under consideration for the treatment of PD patients with psychotic symptoms. Remoxipride is a selective D2 blocker with a preferential affinity to the mesolimbic receptors (Meltzer et al., 1993). Extrapyramidal side-effects are less frequent than in haloperidol but similar to chlorpromazine or thioridazine. In one study examining the efficacy of remoxipride in PD (Lang et al., 1993) nine patients were treated with doses of 75–100 mg/day. Eight out of nine patients improved, but four showed worsening of motor condition. In another open-label pilot study, however, remoxipride was administered to seven PD patients exhibiting thought disorders. Symptoms improved in six patients without deterioration of motor symptoms except in one case (Mendis et al., 1994).

Other atypical neuroleptics in diverse stages of development include melperone and amperozide (Meltzer et al., 1993), sertindole (Borison, 1995), fluperlapine (Scholz and Dichgans, 1985) and trimipramine (Gross et al., 1991), but little clinical information is available.

Serotonin receptors antagonists

Ondansetron

Ondansetron is a serotonin 5-HT3 receptor antagonist used with success in schizophrenic patients at doses of 12–20 mg/day (White et al., 1991). Extrapyramidal side-effects are extremely rare; however, dystonia and akathisia were described in neoplasic patients on chemotherapy receiving ondansetron as antiemetic therapy (Halperin and Murphy, 1992). In a study on seven parkinsonian patients with hallucinations induced by dopaminergic therapy, ondansetron was effective in all, reducing the symptomatology in four of them and abolishing hallucinations in three (Zoldan et al., 1993). Discontinuation of the drug resulted in reappearance of symptoms in a few days. No adverse side-effects were described. In another study by the same author, 16 parkinsonian patients with psychotic symptoms, treated with levodopa, received increasing doses of ondansetron escalated by 4 to 8 mg per week up to a total maintenance dose of 12 to 24 mg/day. Ondansetron did not worsen the motor features or interfere with levodopa efficacy. In addition, hallucinations, paranoid delusions and other psychotic symptoms improved markedly in most of the patients. (Zoldan et al., 1996). Adverse effects were mild consisting of occasional headaches and aggravation of constipation in a few patients. Further studies will be needed to confirm these optimistic findings.

The efficacy of the pure serotonin blockers, such as ondansetron, in the treatment of hallucinosis in PD might be produced by inhibition of corticolimbic serotonergic receptors in drug-induced psychosis. It has been suggested that these receptors are overstimulated by serotonin flushed out of nerve terminals by extraordinarily high cerebrospinal dopamine levels (Klawans, 1988).

Other drugs with antiserotonergic properties that have been used in the treatment of psychosis are ritanserin (Wiesel et al., 1994) which did not demonstrated extrapyramidal side-effects in a study performed in ten patients and setoperone (Ceulemans et al., 1985). Both drugs are selective 5-HT2 receptor antagonists that have been proved effective in the treatment of both negative and positive symptoms of schizophrenia. To our knowledge, there are no clinical trials of these compounds in psychotic PD patients.

The antidepressant mianserin is also a relatively selective 5-HT2 receptor antagonist that was tested in twelve PD patients with drug-induced psychosis. In this study, mianserin was given orally at doses of 20–60 mg/day and the results showed a marked improvement in eight, moderate in two, and no effect in two patients. The parkinsonian motor disability seemed to be decreased slightly in eight of these patients (Ikeguchi and Kuroda, 1995). These results are promising, but they must be reproduced in other controlled trials to ascertain the clinical utility of this compound.

Other drugs

Terguride is a partial D2 agonist, derivative of the ergot agonist lisuride (Akai et al., 1993). Agonism only occurs when the post-synaptic D2 receptors are supersensitive. Thus, loss of striatal neurons, denervation in the striatum and in consequence D2-receptor upregulation, which is the case in PD, induces terguride to have agonist properties. By contrast, in neurons with normal dopamine sensitivity, the drug has an ability to block the D2 receptors. These ambivalent properties might explain in part its antipsychotic effect without worsening motor function (Factor et al., 1995). Animal studies have supported this theory (Akai et al., 1993), but studies in PD patients are not yet available.

References

Albert ML, Feldman RG, Willis A (1974) The "subcortical dementia" of progressive supranuclear palsy. J Neurol Neurosurg Psychiatry 37: 121–130

Algeri S, De Luigi A, De Simoni MG, et al (1988) Multiple and complex effects of buspirone on central dopaminergic system. Psychopharmacol Biochem Behav 29: 823–826

Akai T, Yamaguchi M, Mizuta E, Kuno S (1993) Effects of terguride, a partial D2 agonist, on MPTP-lesioned parkinsonian cynomolgus monkeys. Ann Neurol 33: 507–511

American Psychiatric Association, Committee of Nomenclature and Statistics (1987) Diagnostic and statistical manual of mental disorders, 3rd edn, rev. American Psychiatric Association, Washington DC

Anderson GD, Rebec GV (1988) Clozapine and haloperidol in the amygdaloid complex: differential effects on dopamine transmission with long-term treatment. Biol Psychiatry 23: 497–506

Andrade R, Nicoll RA (1987) Novel anxyolitics discriminate between postsynaptic serotonin receptors mediating different physiological responses on single neurons of the rat hippocampus. Arch Pharmacol 336: 5–10

Baldessarini RJ, Frankenburg FR (1991) Clozapine, a novel antipsychotic drug. N Engl J Med 324: 746–754

Baldessarini RJ, Marsh ER, Kula NS (1992) Interactions of fluoxetine with metabolism of dopamine and serotonin in rat brain regions. Brain Res 579: 152–156

Barbeau A (1969) L-dopa therapy in Parkinson's disease: a critical review of nine years' experience. J Can Med Assoc 101: 791–800

Baxter LR, Schwartz JM, Phelps ME, et al (1989) Reduction of prefrontal glucose metabolism common to three types of depression. Arch Gen Psychiatry 46: 243–250

Beardsley JV, Puletti F (1971) Personality (MMPI) and cognitive (WAIS) changes after levodopa treatment: occurrence in patients with Parkinson's disease. Arch Neurol 25: 145–150

Beasley BL, Nutt JG, Davenport RW, Chase TN (1980) Treatment with tryptophan of levodopa associated psychiatric disturbances. Arch Neurol 37: 155–156

Beck AT, Ward CH, Mandelson M, Mock M, Erbaugh J (1961) An inventory for measuring depression. Arch Gen Psychiatry 4: 561–571

Biggins CA, Boyd JL, Harrop FM, Medeley P, Mindham RHS, Randall JI, Spokes EGS (1992) A controlled, longitudinal study of dementia in Parkinson's disease. J Neurol Neurosurg Psychiatry 55: 566–571

Borison RL (1995) Clinical efficacy of serotonin-dopamine antagonists relative to classic neuroleptics. J Clin Psychopharmacol 15 [Suppl 1]: 24–29

Borison RL, Fields JZ, Diamond BL (1981) Site-specific blockade of dopamine receptors by neuroleptic agents in human brain. Neuropharmacol 20: 1321–1322

Borison RL, Rathiraja AP, Bruce I, et al (1992) Risperidone: clinical safety and efficacy in schizophrenia. Psychopharmacol Bull 28: 213–218

Bouchard RH, Pourcher E, Vincent P (1989) Fluoxetine and extrapyramidal side-effects. Am J Psychiatry 46: 1352–1253

Bowen FP, Kamienny RS, Burns MM, Yahr MD (1975) Parkinsonism: effect of levodopa treatment on concept formation. Neurology 25: 701–704

Brown R, Marsden CD (1986) Visuospatial function in Parkinson's disease. Brain 109: 987–1002

Brown R, Jahnshahi M (1995) Depression in Parkinson's disease: a psychosocial viewpoint. In: Weiner WG, Lang AE (eds) Advances in neurology, vol 65. Behavioural neurology of movement disorders. Raven Press, New York, pp 61–84

Bürki HR, Eichenberger E, Sayers AC, White TG (1975) Clozapine and the dopamine hypothesis of schizophrenia, a critical appraisal. Pharmakopsychiatr Neuropsychopharmakol 8: 115–121

Caparros-Lefebvre D, Pécheux N, Petit V, Duhamel A, Petit H (1995) Which factors predict cognitive decline in Parkinson's disease. J Neurol Neurosurg Psychiatry 58: 51–55

Celesia GG, Wanamaker WM (1972) Psychiatric disturbances in Parkinson's disease. Dis Nerv Syst 33: 577–583

Ceulemans DLS, Gelders YG, Hoppenbrouwers ML, Reyntjens AJ, Janssen PAJ (1985) Effects of serotonin antagonists on schizophrenia: a pilot study with setoperone. Psychopharmacol 85: 329–332

Cheng AVT, Ferrier IN, Morris CM, et al (1991) Cortical serotonin-S2 receptor binding in Lewy body dementia, Alzheimer's and Parkinson's diseases. J Neurol Sci 106: 50–55

Chinaglia G, Landwehrmeyer B, Probst A, Palacios JM (1993) Serotoninergic terminal transporters are differentially affected in Parkinson's disease and progressive supranuclear palsy: an autoradiographic study with (3H)citalopram. Neuroscience 54: 691–699

Chouinard G, Jones B, Remington G, et al (1993) Canadian multi-centre placebo controlled study of fixed doses of risperidone and haloperidol in the treatment of chronic schizophrenic patients. J Clin Psychopharmacol 13: 25–40

Cianchetti C (1985) Dopamine agonists and sleep in man. In: Wauquier A, Gaillard JM, Monti JM, Radulovacki M (eds) Sleep: neurotransmitters and neuromodulators. Raven Press, New York, pp 121–134

Claus A, Bollen J, De Cuyper H, et al (1992) Risperidone versus haloperidol in the treatment of chronic schizophrenic in-patients: a multi-centre, double-blind comparative study. Acta Psychiatr Scand 85: 295–305

Cohen BM, Herschel M, Aoba A (1979) Neuroleptic antimuscarinic and antiadrenergic activity of chlorpromazine, thioridazine and their metabolites. Psychiatry Res 1: 199–208

Cohen S, Chiles J, McNaughton A (1990) Weight gain associated with clozapine. Am J Psychiatry 147: 503–504

Cooper JA, Sagar HJ (1993) Incidental and intentional recall in Parkinson's disease: an account based on diminished attentional resources. J Clin Exp Neuropsychol 15: 713–731

Cooper JA, Sagar HJ, Doherty SM, et al (1992) Different effects of dopaminergic and anticholinergic therapy on memory performance in Parkinson's disease. Brain 115: 1701–1725

Cowen PJ (1993) Serotonin receptor subtypes in depression: evidence from studies in neuroendocrine regulation. Clin Neuropharmacol 16 [Suppl 3]: 6–18

D'Amato RJ, Zweig RM, Whitehouse PJ, et al (1987) Aminergic systems in Alzheimer's disease and Parkinson's disease. Ann Neurol 22: 229–236

De Vos RA, Jansen EN, Stam FC, Ravid R, Swab DF (1995) "Lewy body disease": clinico-pathological correlations in 18 consecutive cases of Parkinson's disease with and without dementia. Clin Neurol Neurosurg 97: 12–22

Dogali M, Fazzini E, Kolodny E, et al (1995) Stereotactic ventral pallidotomy for Parkinson's disease. Neurology 45: 753–761

Dogali M, Sterio D, Fazzini E, et al (1996) Effects of posteroventral pallidotomy in Parkinson's disease. In: Battistin L, Scarlato G, Caraceni T, Ruggieri S (eds) Advances in neurology, vol 69. Parkinson's disease. Lippincot-Raven, Philadelphia, pp 585–590

Donelli EF, Chase TN (1972) Intellectual and memory function in parkinsonian and non-parkinsonian patients treated with l-dopa. Dis Nerv Syst 34: 119–123

Dubois B, Danze F, Pilon B, et al (1987) Cholinergic-dependent deficits in Parkinson's disease. Ann Neurol 22: 26–30

Duncombe ME, Bradshaw JL, Iansek R, Phillips JG (1994) Parkinsonian patients without dementia or depression do not suffer from bradyphrenia as indexed by performance in mental rotation tasks with and without advance information. Neuropsychologia 32: 1383–1396

Eison AS, Eison MS, Torrente JR, Wright RN, Yocca FD (1990) Nefazodone: preclinical pharmacology of a new antidepressant. Psychopharmacol Bull 26: 311–315

Factor SA, Brown D (1992) Clozapine prevents recurrence of psychosis in Parkinson's disease. Mov Disord 7: 125–131

Factor SA, McAlarney T, Sánchez-Ramos JR, Weiner WJ (1990) Sleep disorders and sleep effect in Parkinson's disease. Mov Disord 5: 280–285

Factor SA, Brown D, Molho ES, Podskalny GD (1994) Clozapine: a 2-year open trial in Parkinson's disease patients with psychosis. Neurology 44: 544–546

Factor SA, Molho ES, Podskalny GD, Brown D (1995) Parkinson's disease: drug-induced psychiatric states. In: Weiner WJ, Lang AE (eds) Advances in neurology, vol 65. Raven Press, New York, pp 115–138

Fahn S, Libsch LR, Cutler RW (1971) Monoamines in the human neostriatum: topographic distribution in normals and in Parkinson's disease in their role in akinesia, rigidity, chorea and tremor. J Neurol Sci 14: 427–455

Farde L, Nordström A-L (1992) PET analysis indicates typical central dopamine receptor occupancy in clozapine-treated patients. Br J Psychiatry 160 [Suppl 17]: 30–33

Farde L, Wiesel FA, Nordström A-L, Sedvall G (1989) D1- and D2-dopamine receptor occupancy during treatment with conventional and atypical neuroleptics. Psychopharmacol 99: 28–31

Farina E, Cappa SF, Polimeni M, Magni E, Canesi M, Zechinelli A, Scarlato G, Mariani C (1994) Frontal dysfunction in early Parkinson's disease. Acta Neurol Scand 90: 34–38

Feighner JP, Pambakian R, Fowler RC, et al (1989) A comparison of nefazodone, imipramine and placebo in patients with severe to moderate depression. Psychopharmacol Bull 25: 219–221

Feighner JP, Boyer WF, Tyler DL, Neborsky RJ (1990) Adverse consequences of fluoxetine-MAOI combination therapy. J Clin Psychiatry 51: 222–225

Folstein MF, Folstein SE, McHugh PR (1975) "Mini-Mental State", a practical method for grading the cognitive state of patients for the clinician. J Psychiatr Res 12: 189–198

Fontaine R (1993) Novel serotonergic mechanisms and clinical experience with nefazodone. Clin Neuropharmacol 16 [Suppl 3]: 45–50

Ford B, Lynch T, Greene P (1994) Risperidone in Parkinson's disease. Lancet 344: 681

Fram D, Murphy D, Goodwin F, et al (1970) L-Dopa and sleep in depressed patients. Psychophysiology 7: 316–317

Friedman JH (1991) The management of the levodopa psychoses. Clin Neuropharmacol 14: 283–295

Friedman JH, Lannon MC (1989) Clozapine in the treatment of psychosis in Parkinson's disease. Neurology 39: 1219–1221

Fungfeld EW, Baggen M, Nedwidek P, et al (1989) Double blind study with phosphatidylserine in parkinsonian patients with senile dementia of Alzheimer's type (SDAT). Prog Clin Biol Res 317: 1235–1246

Gedye A (1991) Buspirone alone or with serotonergic diet reduced aggression in a developmentally disabled adult. Biol Psychiatry 30: 88–91

Gerlach J, Peacock L (1994) Motor and mental side-effects of clozapine. J Clin Psychiatry 55 S B: S107–S109

German DC, Menaye K, Smith WK, Woodward DJ, Saper CB (1989) Midbrain dopaminergic cell loss in Parkinson's disease: computer visualization. Ann Neurol 26: 507–514

Gerson SL (1993) Clozapine- deciphering the risks. N Engl J Med 329: 204–205

Goetz CG, Stebbins GT (1993) Risk factors for nursing home placement in advanced Parkinson's disease. Neurology 43: 2227–2229

Gómez Arévalo G, Gershanik OS (1993) Modulatory effect of clozapine on levodopa response in Parkinson's disease: a preliminary study. Mov Disord 8: 349–354

Goodwin FK (1971) Psychiatric side-effects of levodopa in man. JAMA 218: 1915–1920

Gotham AM, Brown RG, Marsden CD (1986) Depression in Parkinson's disease: a quantitative and qualitative analysis. J Neurol Neurosurg Psychiatry 49: 381–389

Gross G, Xin X, Gastpar M (1991) Trimipramine: pharmacological reevaluation and comparison with clozapine. Neuropharmacol 30: 1159–1166

Haring C, Neudorfer C, Schwitzer J, et al (1994) EEG alterations in patients treated with clozapine in relation to plasma levels. Psychopharmacol 114: 97–100

Halgin R, Riklin M, Misiak H (1977) Levodopa, parkinsonism, and recent memory. J Nerv Ment Dis 164: 268–272

Halliday GM, Blumbergs PC, Cotton RGH, et al (1990) Loss of brainstem serotonin- and substance P- containing neurons in Parkinson's disease. Brain Res 510: 104–107

Halperin JR, Murphy B (1992) Extrapyramidal reaction to ondansetron. Cancer 69: 1275

Hamilton M (1960) A rating scale for depression. J Neurol Neurosurg Psychiatry 23: 56–62

Harnel AR, Riklan M (1975) Cognitive and perceptual effects of long range l-dopa therapy in parkinsonism. J Clin Psychol 31: 321–333

Heinrich K, Klieser E, Lehmann E, Kinzler E (1991) Experimental comparison of the efficacy and compatibility of risperidone and clozapine in acute schizophrenia. In: Kane JM (ed) Risperidone: major progress in antipsychotic treatment. Oxford Clinical Communications, Oxford, pp 37–39

Hietanen M, Teräväinen H (1988) The effect of age of disease onset on neuro-psychological performance in Parkinson's disease. J Neurol Neurosurg Psychiatry 51: 244–249

Huber SJ, Shulman HG, Paulson GW, Shuttleworth EC (1987) Fluctuations in plasma dopamine level impair memory in Parkinson's disease. Neurology 37: 1371–1375

Huber SJ, Shulman HG, Paulson GW, Shuttleworth EC (1989) Dose-dependent memory impairment in Parkinson's disease. Neurology 39: 438–440

Hughes AJ, Daniel SE, Kilford L, et al (1992) Accuracy of diagnosis of idiopathic Parkinson's disease: a clinicopathological study of 100 cases. J Neurol Neurosurg Psychiatry 55: 181–184

Hurwitz TA, Calne DB, Waterman K (1988) Treatment of dopaminomimetic psychosis in Parkinson's disease with electroconvulsive therapy. Can J Neurol Sci 15: 32–34

Idänpään-Heikkilä J, Alhava E, Olkinuora M, Palva I (1975) Clozapine and agranulocytosis. Lancet ii: 611

Ikeguchi K, Kuroda A (1995) Mianserin treatment of patients with psychosis induced by antiparkinsonian drugs. Eur Arch Psychiatry Clin Neurosci 244: 320–324

Iversen LL (1993) The D4 and schizophrenia. Nature 365: 393

Jankovic J, McDermott M, Carter J, et al (1990) Variable expression of Parkinson's disease: a base-line analysis of the DATATOP cohort. Neurology 40: 1529–1534

Jansen PAJ (1987) The development of new antipsychotic drugs: towards a new strategy in the management of chronic psychoses. J Drug Ther Res 12: 324–328

Jansen Steur ENH (1993) Increase of Parkinson disability after fluoxetine medication. Neurology 43: 211–213

Jansen ENH (1994) Clozapine in the treatment of tremor in Parkinson's disease. Acta Neurol Scand 89: 262–265

Javoy-Agid F, Taquet H, Ploska A, et al (1981) Distribution of catecholamines in the ventral mesencephalon of the human brain, with special reference to Parkinson's disease. J Neurochem 36: 2101–2105

Jiménez-Jiménez FJ, Tejeiro J, Martínez-Junquera G, et al (1994) Parkinsonism exacerbated by paroxetine. Neurology 44: 2406

Juul-Povlsen U, Noring U, Fog R, Gerlach J (1985) Tolerability and therapeutic effect of clozapine: a retrospective investigation of 216 patients, treated with clozapine for up to 12 years. Acta Psychiatr Scand 71: 176–185

Kafi S, Gaillard J (1976) Brain dopamine receptors and sleep in the rat: effects of stimulation and blockade. Eur J Pharmacol 38: 357–363

Karper LP, Salloway SP, Seibyl JP, Krystal JH (1992) Prolonged post-ictal encephalopathy in two patients with clozapine induced seizures. J Neuropsychiatry Clin Neurosci 4: 454–457

Kazee AM, Han LY (1995) Cortical Lewy bodies in Alzheimer's disease. Arch Path Lab Med 119: 448–453

Kieburtz K, McDermott M, Como P, et al (1994) The effect of deprenyl and tocopherol on cognitive performance in early untreated Parkinson's disease. Neurology 44: 1756–1759

Klawans HL (1988) Psychiatric side-effects during the treatment of Parkinson's disease. J Neural Transm 27 [Suppl]: 117–122

Koller WC (1984) Disturbance of recent memory function in parkinsonian patients on anticholinergic therapy. Cortex 20: 307–311

Krupp P, Barnes P (1992) Clozapine-associated agranulocytosis: risk and aetiology. Br J Psychiatry 160 [Suppl 17]: 38–40

Lang AE, Johnson K (1987) Akathisia in idiopathic Parkinson's disease. Neurology 37: 477–481

Lang AE, Sandor P, Duff J (1993) Remoxipride in Parkinson's disease: differential response in patients with dyskinesias/fluctuations vs. psychosis. Ann Neurol 34: 301–302 (Abstract)

Launer M (1992) Diarrhoea during treatment with clozapine. Br Med J 305: 1160

Levin BE, Llabre MM, Reisman S (1991) A retrospective analysis of the effects of anticholinergic medication on memory performance in Parkinson's disease. J Neuropsychiatry Clin Neurosci 3: 412–416

Lew MF, Waters CH (1993) Clozapine treatment of parkinsonism with psychosis. J Am Geriatr Soc 41: 669–671

Lipinski JF Jr, Mallya G, Zimmerman, Pope HG Jr (1989) Fluoxetine-induced akathisia: clinical and theoretical implications. J Clin Psychiatry 50: 339–342

Lipper S (1976) Psychosis in patients on bromocriptine and levodopa with carbidopa. Lancet ii: 571–572

Litvan I, Mohr E, Williams J, Fedio P, Chase TN (1991) Differential memory and executive functions in demented patients with Parkinson's and Alzheimer's disease. J Neurol Neurosurg Psychiatry 54: 25–29

Livingston MG (1994) Risperidone. Lancet 343: 457–460

Lozano AM, Lang AE, Gálvez-Jiménez N, et al (1995) Effect of GPi pallidotomy on motor function in Parkinson's disease. Lancet 346: 1383–1387

Marsden CD, Parkes JD, Quinn N (1982) Fluctuations of disability in Parkinson's disease: clinical aspects. In: Marsden CD, Fahn S (eds) Movement disorders. Butterworth Scientific, London, pp 96–122

Martí MJ, Tolosa E (1991) Demencia en la enfermedad de Parkinson y en la enfermedad de Alzheimer: similitudes y diferencias. In: Esquerda JE, Gallego R, Gual A, Ramírez G, Rubia F (eds) Neurotransmisión y plasticidad sináptica. Espaxs, Barcelona, pp 293–309

Mayberg HS, Solomon DH (1995) Depression in Parkinson's disease: a biochemical and organic viewpoint. In: Weiner WJ, Lang AE (eds) Advances in neurology, vol 65. Behavioral neurology of movement disorders. Raven Press, New York, pp 49–60

Mayberg HS, Starkstein SE, Sadzot B, et al (1990) Selective hypometabolism in the inferior frontal lobe in depressed patients with Parkinson's disease. Ann Neurol 28: 57–64

Mayeux R, Stern Y, Rosen J, Leventhal J (1981) Depression, intellectual impairment, and Parkinson's disease. Neurology 31: 645–650

Mayeux R, Stern Y, Cote L, Williams JBW (1984) Altered serotonin metabolism in depressed patients with Parkinson's disease. Neurology 34: 642–646

Mayeux R, Stern Y, Williams JBW, et al (1986) Clinical and biochemical features of depression in Parkinson's disease. Am J Psychiatry 143: 756–759

Mayeux R, Denaro J, Hemenegildo N, Marder K, Tang M-X, Cote LJ, Stern Y (1992) A population-based investigation of Parkinson's disease with and without dementia. Arch Neurol 49: 492–497

McCance-Katz EF, Marek KL, Price LH (1992) Serotonergic dysfunction in depression associated with Parkinson's disease. Neurology 42: 1813–1814

McDermott HP, Jankovic J, Parkinson Study Group (1992) Factors predictive of the need for levodopa therapy in early, untreated Parkinson's disease. Neurology 42 [Suppl 3]: 441

Meco G, Alessandria A, Binfati V, Giustini P (1994) Risperidone for hallucinations in levodopa-treated Parkinson's disease patients. Lancet 343: 1370–1371

Megens AAHP, Awouters FHL, Schotte A, et al (1994) Survey on the pharmacodynamics of the new antipsychotic risperidone. Psychopharmacol 114: 9–23

Meier MJ, Martin WE (1970) Intellectual changes associated with levodopa therapy. JAMA 213: 456–466

Melamed E, Zoldan J, Friedberg G, Goldberg-Stein H (1993) Is hallucinosis in Parkinson's disease due to central serotonergic hyperactivity? Mov Disord 8: 406–407

Meltzer HY (1993) New drugs for the treatment of schizophrenia. Psychiatr Clin North Am 16: 365–385

Meltzer HY, Koenig JL, Nash JF, Gudelsky GA (1989a) Melperone and clozapine: neuroendocrine effects of typical neuroleptic drugs. Acta Psychiatr Scand 352 [Suppl]: 24–29

Meltzer HY, Matsubara S, Lee J-C (1989b) Classification of typical and atypical antipsychotic drugs on the basis of dopamine D1, D2 and serotonin-2 pKi values. J Pharmacol Exp Ther 251: 238–246

Meltzer HY, Kennedy J, Dai J, Parsa M, Riley D (1995) Plasma clozapine levels and the treatment of L-DOPA-induced psychosis in Parkinson's disease. A high potency effect of clozapine. Neuropsychopharmacol 12: 39–45

Mendis T, Mohr E, George A, et al (1994) Symptomatic relief from treatment-induced psychosis in Parkinson's disease: an open-label pilot study with remoxipride. Mov Disord 9: 197–200

Mindham RHS (1970) Psychiatric symptoms in parkinsonism. J Neurol Neurosurg Psychiatry 33: 181–191

Mohr E, Mendis T, Grimes DJ (1995) Late cognitive changes in Parkinson's disease with emphasis on dementia. In: Weiner WJ, Lang AE (eds) Advances in neurology, vol 65. Behavioral neurology of movement disorders. Raven Press, New York, pp 97–113

Molinari SP, Kaminski R, Di Rocco A, Yahr MD (1995) The use of famotidine in the treatment of Parkinson's disease: a pilot study. J Neural Transm [P-D Sect] 9: 243–247

Montastruc JL, Fabre N, Blin O, et al (1994) Does fluoxetine aggravate Parkinson's disease? A pilot prospective trial. Mov Disord 9 [Suppl 1]: 99

Moore NA, Axton MS (1988) Production of climbing behaviour in mice requires both D1 and D2 receptor activation. Psychopharmacol 94: 263–266

Moore NA, Tye NC, Axton MS, Risius FC (1992) The behavioural pharmacology of olanzapine, a novel "atypical" antipsychotic agent. J Pharmacol Exp Ther 262: 545–551

Moore S, Kenyon P (1994) Atypical antipsychotics, clozapine and sulpiride do not antagonise amphetamine-induced stereotyped locomotion. Psychopharmacol 114: 123–130

Moskovitz C, Moses H, Klawans H (1978) Levodopa induced psychosis: a kindling phenomena. Am J Psychiatry 135: 6–10

Müller T, Becker T, Fritze J (1988) Neuroleptic malignant syndrome after clozapine plus carbamacepine. Lancet ii: 1500

Naber D, Holzbach R, Perro C, Hippius H (1992) Clinical management of clozapine patients in relation to efficacy and side-effects. Br J Psychiatry 160 [Suppl 17]: 54–59

Narabayashi H (1996) Does stereotactic treatment of Parkinson's disease slow the progression of the disease? In: Battistin L, Scarlato G, Caraceni T, Ruggieri S (eds) Advances in neurology, vol 69. Parkinson's disease. Lippincot-Raven, Philadelphia, pp 557–562

Nausieda PA (1992) Sleep in Parkinson's disease. In: Koller WC (ed) Handbook of Parkinson's disease. Marcel Dekker, New York, pp 451–467

Nausieda PA, Weiner WJ, Kaplan LR, Weber S, Klawans HL (1982) Sleep disruption in the course of chronic levodopa therapy: an early feature of the levodopa psychosis. Clin Neuropharmacol 5: 183–194

Nausieda P, Tanner C, Klawans H (1983) Serotonergically active agents in the treatment of the levodopa induced psychosis. In: Fahn S, Calne DB, Shoulson I (eds) Advances in neurology, vol 37. Experimental therapeutics of movement disorders. Raven Press, New York, pp 23–32

Nausieda PA, Leo GJ, Chesney D (1994) Comparison of conventional and Sinemet CR on the sleep of parkinsonian patients. Neurology 44 [Suppl 2]: A219

Nutt JG, Gancher ST, Woodward WR (1988) Does inhibitory action of levodopa contribute to motor fluctuations? Neurology 38: 1553–1557

Ostergaard K, Dupond E (1988) Clozapine treatment of drug-induced psychotic symptoms in late stages of Parkinson's disease. Acta Neurol Scand 78: 349–350

Palazzini E, Soliveri P, Filippini G, et al (1995) Progression of motor and cognitive impairment in Parkinson's disease. J Neurol 242: 535–540

Parkinson J (1817) An essay on the shaking palsy. Whittingham and Rowland for Sherwood, Neely, and Jones, London

Parkinson Study Group (1993) Effect of deprenyl and tocopherol on the progression of disability in early Parkinson's disease. N Engl J Med 328: 176–183

Pfeiffer RF, Kang J, Graber B, Hofman R, Wilson J (1990) Clozapine for psychosis in Parkinson's disease. Mov Disord 5: 239–242

Pilowski LS, Costa DC, Elli PJ, et al (1992) Clozapine, single photon emission tomography, and the D2 dopamine receptor blockade hypothesis of schizophrenia. Lancet 340: 199–202

Pirozzolo FJ, Swihart AA, Rey GJ, Mahurin R, Jankovic J (1993) Cognitive impairments associated with Parkinson's disease and other movement disorders. In: Jankovic J, Tolosa E (eds) Parkinson's disease and movement disorders. Williams & Wilkins, Baltimore, pp 493–510

Pope HG Jr, Cole JO, Choras PT, Fulwiler CE (1986) Apparent neuroleptic malignant syndrome with clozapine and lithium. J Nerv Ment Dis 174: 493–495

Rajput AH, Pahwa R, Pahwa P (1992) Mode of onset and prognosis in Parkinson's disease. Neurology 42 [Suppl 3]: 419

Richelson E, Nelson A (1984) Antagonism by neuroleptics of neurotransmitter receptors of normal human brain in vivo. Eur J Pharmacol 103: 197–204

Ring H, Bench CJ, Trimble MR, Brooks DJ, Frackowiak RS, Dolan RJ (1994) Depression in Parkinson's disease. Br J Psychiatry 165: 333–339

Rinne UK (1983) Problems associated with long term levodopa treatment of Parkinson's disease. Acta Neurol Scand 95: 19–26

Robinson RG (1988) Post-stroke depression and lesion location. Stroke 19: 125

Rubinstein M, Gershanik OS, Stefano FJE (1988) Different roles of D1 and D2 dopamine receptors involved in locomotor activity in supersensitive mice. Eur J Pharmacol 148: 419–426

Rupniak NMJ, Hall MD, Mann S, et al (1985) Chronic treatment of clozapine, unlike haloperidol, does not induce changes in D-2 receptor function in the rat. Biochem Pharmacol 34: 2755–2763

Sadeh MS, Braham J, Modan M (1982) Effects of anticholinergic drugs on memory in Parkinson's disease. Arch Neurol 39: 666–667

Sano M, Stern Y, Marder K, Mayeux R (1990) A controlled trial of piracetam in intellectually impaired patients with Parkinson's disease. Mov Disord 5: 230–234

Santamaria J, Tolosa E (1992) Clinical subtypes of Parkinson's disease and depression. In: Huber SG, Cummings JL (eds) Parkinson's disease. Neurobehavioral aspects. Oxford University Press, New York, pp 217–228

Santamaria J, Tolosa E, Vallés A (1986) Parkinson's disease with depression: a possible subgroup of idiopathic parkinsonism. Neurology 36: 1130–1133

Scharf B, Moskowitz C, Moses H, Luptom M, Klawans H (1978) Dream phenomena induced by chronic levodopa therapy. J Neural Transm 43: 143–151

Scholz E, Dichgans J (1985) Treatment of drug-induced exogenous psychosis in parkinsonism with clozapine and fluperlapine. Eur Arch Psychiatry 235: 60–64

Seeman P (1992) Dopamine receptor sequences, therapeutic levels of neuroleptics occupy D2 receptors, clozapine occupies D4. Neuropsychopharmacol 7: 261–284

Seeman P, Guan HC, Van Tol HHM (1993) Dopamine D4 receptors elevated in schizophrenia. Nature 365: 441–445

Shaikh S, Coller D, Kerwin RW, et al (1993) Dopamine D4 receptor subtypes and response to clozapine. Lancet 341: 116

Shaw KM, Lees AJ, Stern GM (1980) The impact of treatment with levodopa on Parkinson's disease. Q J Med 49: 283–293

Sibley DR, Monsma FJ (1992) Molecular biology of dopamine receptors. TIPS 13: 61–69

Simpson GM, Cooper TA (1978) Clozapine plasma levels and convulsions. Am J Psychiatry 135: 99–100

Spencer DD, Robbins RJ, Naftolin F, et al (1992) Unilateral transplantation of human fetal mesencephalic tissue into the caudate nucleus of patients with Parkinson's disease. N Engl J Med 327: 1541–1548

Starkstein SE, Preziosi TJ, Berthier ML, et al (1989) Depression and cognitive impairment in Parkinson's disease. Brain 112: 1141–1153

Starkstein SE, Bolduc PL, Mayberg HS, Preziosi TJ, Robinson RG (1990) Cognitive impairment and depression in Parkinson's disease: a follow-up study. J Neurol Neurosurg Psychiatry 53: 597–602

Stern Y, Marder K, Tang MX, Mayeux R (1993) Antecedent clinical features associated with dementia in Parkinson's disease. Neurology 50: 1192–1196

Stein MB, Heuser IJ, Juncos JL, Uhde TW (1990) Anxiety disorders in patients with Parkinson's disease. Am J Psychiatry 147: 217–220

Tasker RR, DeCarvalho GC, Li CS, Kestle JRW (1996) Does thalamotomy alter the course of Parkinson's disease? In: Battistin L, Scarlato G, Caraceni T, Ruggieri S (eds) Advances in neurology, vol 69. Parkinson's disease. Lippincot-Raven, Philadelphia, pp 563–584

Thomas DR, Nelson DR, Johnson AM (1987) Biochemical effects of the antidepressant paroxetine, a specific 5-hydroxytryptamine uptake inhibitor. Psychopharmacology (Berlin) 93: 193–200

Thomas P, Goudeman M (1992) Seizures with low doses of clozapine. Am J Psychiatry 149: 138–139

Tison F, Dartigues JF, Auriacombe S, Letenneur L, Boller F, Alpérovitcj A (1995) Dementia in Parkinson's disease: a population-based study in ambulatory and instutionalized individuals. Neurology 45: 705–708

Tolosa E, Valldeoriola F, Martí MJ (1994) New and emerging strategies for improving levodopa treatment. Neurology 44 [Suppl 6]: 35–44

Tröster AI, Paolo AM, Lyons KE, Glatt SL, Hubble JP, Koller WC (1995) The influence of depression on cognition in Parkinson's disease: a pattern of impairment distinguishable from Alzheimer's disease. Neurology 45: 672–676

Van Beijsterveldt LEC, Geerts RJF, Leysen JE, et al (1994) Regional brain distribution of risperidone and its active metabolite 9-hydroxy-risperidone in the rat. Psychopharmacol 114: 53–62

Van Herwaarden G, Berger HJC, Horstink MWIM (1993) Short-term memory in Parkinson's disease after withdrawel of long-term anticholinergic therapy. Clin Neuropharmacol 16: 438–443

Van Tol HHM, Bunzow JR, Guan HC, et al (1991) Cloning of the gene for a human dopamine D4 receptor with high affinity for the antipsychotic clozapine. Nature 350: 610–619

Walker JM, Matsumoto RR, Bowen WD, Gans DL, Jones KD, Walker FO (1988) Evidence for a role of haloperidol-sensitive sigma "opiate" receptors in the motor effects of antipsychotic drugs. Neurology 38: 961–965

Waters CH (1994) Fluoxetine and selegiline: lack of significant interaction. Can J Neurol Sci 21: 259–261

White A, Corn TH, Feetham, Faulconbridge C (1991) Ondansetron in treatment of schizophrenia. Lancet 337: 1173

Wiesel FA, Nordström AL, Farde L, Eriksson B (1994) An open clinical and biochemical study of ritanserin in acute patients with schizophrenia. Psychopharmacol 114: 31–38

Wilson JA, Smith RG (1989) The prevalence and aetiology of long term L-dopa side-effects in elderly parkinsonian patients. Age Ageing 18: 11–16

Wolters EC, Hurwitz TA, Mak E, et al (1990) Clozapine in the treatment of parkinsonian patients with dopaminomimetic psychosis. Neurology 40: 832–834

Yahr MD, Duvoisin RS, Schear MJ, Barrett RE, Hoehn MM (1969) Treatment of parkinsonism with levodopa. Arch Neurol 21: 343–354

Yaryura-Tobias JA, Diamond B, Merlis S (1970) Psychiatric manifestations of levodopa. Dis Nerv Syst 31: 60–63

Zoldan J, Friedberg G, Goldberg-Stern, Melamed E (1993) Ondansentron for hallucinosis in advanced Parkinson's disease. Lancet 341: 562–563

Zoldan J, Friedberg G, Weizman A, Melamed E (1996) Ondansetron, a 5-HT3 antagonist for visual hallucinations and paranoid delusional disorder associated with chronic L-DOPA therapy in advanced Parkinson's disease. In: Battistin L, Scarlato G, Caraceni T, Ruggieri S (eds) Advances in neurology, vol 69. Parkinson's disease. Lippincott-Raven, Philadelphia, pp 541–544

Authors' address: Dr. E. Tolosa, Neurology Service, Hospital Clínic i Provincial, C/Villarroel 170, E-08036 Barcelona, Spain

Springer Neurology

P. Riederer, D. B. Calne, R. Horowski,
Y. Mizuno, W. Poewe, M. B. H. Youdim (eds.)

Advances in Research on Neurodegeneration
Volume 5

1997. 45 figures. VIII, 215 pages.
Cloth DM 198,–, öS 1386,–
ISBN 3-211-82933-4
Special edition of "Journal of Neural Transmission, Supplement 50, 1997"
(Soft cover edition of Supplement 50 only available for subscribers to "Journal of Neural Transmission")

The "International Winter Conferences on Neurodegeneration" have become an established forum to discuss various aspects of basic and clinical topics related to the underlying mechanisms of neurodegenerative disorders. This volume focuses on brain imaging, endogenous and exogenous neurotoxins, programmed cell death, apoptosis and necrosis, and immunoinflammatory mechanisms, infective diseases causing neurological disorders. These topics have been reviewed by invited experts and the articles give an up-to-date reflection of the state of the art in these research fields.

Y. Mizuno, M. B. H. Youdim, D. B. Calne,
R. Horowski, W. Poewe, P. Riederer (eds.)

Advances in Research on Neurodegeneration
Volume 3 & 4

1997. 46 figures. VIII, 280 pages.
Cloth DM 215,–, öS 1505,–
ISBN 3-211-82935-0
Special edition of "Journal of Neural Transmission, Supplement 49, 1997"
(Soft cover edition of Supplement 49 only available for subscribers to "Journal of Neural Transmission")

The first part of the book focuses on disease models and mechanisms. The areas discussed include Alzheimer's disease, Parkinson's disease, glial and neuronal death, and demyelination/remyelination. The second part concentrates on the molecular biology of neurodegeneration. The topics include molecular genetics of neurological disorders, molecular biology of recognition sites, apoptosis, and neuroimmunology and multiple sclerosis. Leading experts have been invited to give state of the art presentations including their own recent data.

SpringerWien NewYork

Sachsenplatz 4-6, P.O.Box 89, A-1201 Wien, Fax +43-1-330 24 26, e-mail: order@springer.at, Internet: http://www.springer.at
New York, NY 10010, 175 Fifth Avenue • D-14197 Berlin, Heidelberger Platz 3 • Tokyo 113, 3-13, Hongo 3-chome, Bunkyo-ku

SpringerNeurology

W. Kuhn, P. Kraus, H. Przuntek (eds.)

Deprenyl – Past and Future

1996. 16 figures. IX, 112 pages.
Cloth DM 130,–, öS 910,–
ISBN 3-211-82948-2
Special edition of "Journal of Neural Transmission", Supplement 48
(Soft cover edition of Supplement 48 only available for subscribers to "Journal of Neural Transmission")

The clinical effect of L-Deprenyl was originally explained on the basis of irreversible and selective MAO-B inhibition and subsequent enhancement of dopaminergic neurotransmission. In recent years new experimental data have challenged this concept. In vitro and in vivo studies are suggesting that L-Deprenyl may have neuroprotective and/or neuroregenerative properties, too. Furthermore, controversial data of recently finished long-term clinical studies have brought forward an new discussion both on the clinical impact and the possible mode of action of L-Deprenyl in Parkinson's Disease and various other neurological and psychiatric disorders. This volume provides a forum for intensive discussions on the biochemical, pharmacological and clinical aspects of Parkinson's Disease.

K. A. Jellinger, M. Windisch (eds.)

New Trends in the Diagnosis and Therapy of Non-Alzheimer's Dementia

1996. 61 partly coloured figures. VIII, 288 pages.
Soft cover DM 190,–, öS 1330,–
Reduced price for subscribers to "Journal of Neural Transmission":
Soft cover DM 171,–, öS 1197,–. ISBN 3-211-82823-0
Journal of Neural Transmission, Supplement 47

This volume gives an overview of the present state of art on the classification, neuropathology, clinical presentation, neuropsychology, diagnosis, neuroimaging and therapeutic possibilities in non-Alzheimer's dementias, an increasingly important group of CNS diseases, which account for 7 to 30% of dementing disorders in adults and aged subjects, and thus, represent the second most frequent cause of dementia after Alzheimer's disease. The monograph provides the newest information for neurologists, psychiatrists, dementia research workers, dementia clinicians, neuropathologists, neurobiologists, and practicing physicians.

SpringerWienNewYork

Sachsenplatz 4-6, P.O.Box 89, A-1201 Wien, Fax +43-1-330 24 26, e-mail: order@springer.at, Internet: http://www.springer.at
New York, NY 10010, 175 Fifth Avenue • D-14197 Berlin, Heidelberger Platz 3 • Tokyo 113, 3-13, Hongo 3-chome, Bunkyo-ku

Parkinson's Disease: Experimental Models and Therapy

1995. 121 figures. XI, 466 pages.
Soft cover DM 240,–, öS 1680,–
Reduced price for subscribers to "Journal of Neural Transmission":
Soft cover DM 216,–, öS 1512,–. ISBN 3-211-82749-8
Journal of Neural Transmission, Supplement 46

Current research on Parkinson's disease is aimed at the goal of determining the underlying cause of this terrible disease and of developing adequate treatment strategies to deal with it. This volume focuses on models that mirror the progression of the symptoms of Parkinson's disease (iron, MPTP, 6-hydroxydopamine, "TaClo", etc.) while other topics are the evaluation of oxidative stress, calcium, excitotoxicity, nitric oxide, or nerve growth factors as possible pathophysiological candidates or causal parameters. Further topics are the interplay between exogenous and endogenous toxins, the potential of brain imaging by PET, MRI and SPECT, as well as promising therapeutic drug strategies. This volume represents a comprehensive survey of the state of the art for neurologists, biochemists, neuropharmacologists and toxicologists.

Old and New Dopamine Agonists in Parkinson's Disease

1995. 73 figures. VIII, 321 pages.
Soft cover DM 196,–, öS 1386,–
Reduced price for subscribers to "Journal of Neural Transmission":
Soft cover DM 176,40, öS 1247,40. ISBN 3-211-82717-X
Journal of Neural Transmission, Supplement 45

This book provides a comprehensive overview of the basic and clinical neuropharmacology of dopamine agonists and the rationale for their employment in PD. The authors have compiled an up-to-date guide, covering such topics as the pathophysiology of dopaminergic systems and the neuro-biochemistry of dopaminergic receptors, the clinical use of old and new dopamine agonists, both in the first-time treatment of PD patients and for reducing motor fluctuations in levodopa-treated ones, and the possible role of dopamine agonists as neuroprotective agents. Particular emphasis has been placed on apomorphine, an old dopamine agonist that has recently recaptured neurologists' interest for its use in both diagnostic use and therapeutic management of advanced parkinsonian patients. Articles discussing the results of ongoing clinical studies of newly developed dopamine agonists and the potential use of dopamine agonists, both new and old, as neuroprotectors should be of particular interest to the reader. The work is an exhaustive up-to-date compendium that assembles the entire spectrum of current basic and clinical research on dopaminergic systems and dopamine agonists in Parkinson's disease into a single authoritative source.

 SpringerWien NewYork

Sachsenplatz 4-6, P.O.Box 89, A-1201 Wien, Fax +43-1-330 24 26, e-mail: order@springer.at, Internet: http://www.springer.at
New York, NY 10010, 175 Fifth Avenue • D-14197 Berlin, Heidelberger Platz 3 • Tokyo 113, 3-13, Hongo 3-chome, Bunkyo-ku

Springer-Verlag
and the Environment

WE AT SPRINGER-VERLAG FIRMLY BELIEVE THAT AN international science publisher has a special obligation to the environment, and our corporate policies consistently reflect this conviction.

WE ALSO EXPECT OUR BUSINESS PARTNERS – PRINTERS, paper mills, packaging manufacturers, etc. – to commit themselves to using environmentally friendly materials and production processes.

THE PAPER IN THIS BOOK IS MADE FROM NO-CHLORINE pulp and is acid free, in conformance with international standards for paper permanency.